INORGANIC THERMODYNAMICS

An Introduction
to the Thermodynamics
of Inorganic
and Organometallic
Compounds

INORGANIC THERMODYNAMICS

An Introduction to the Thermodynamics of Inorganic and Organometallic Compounds

S. J. ASHCROFT
Lecturer in Chemical Engineering
University of Exeter

G. BEECH
Senior Lecturer in Inorganic Chemistry
Wolverhampton Polytechnic

VAN NOSTRAND REINHOLD COMPANY LTD.
New York · Cincinnati · Toronto · London · Melbourne

ISBN 0442 0352 8 cl
ISBN 0442 0353 6 pb

Published by Van Nostrand Reinhold Company Ltd, 25–28 Buckingham Gate, London SW1E 6LQ.

Text set in 10/12 pt. IBM Press Roman, printed by photolithography, and bound in Great Britain at The Pitman Press, Bath.

Contents

Preface

Many of the textbooks relating to thermodynamics or bonding in inorganic chemistry are of a specialist type, dealing with selected areas in depth, and are unlikely to be read by undergraduate students.

We felt that there was a need for a modern up-to-date coverage of that part of thermodynamics which is relevant to bonding in inorganic compounds. As a result, we have written this selective, but introductory treatment, aimed at the undergraduate level. In addition to the more conventional topics, we have included sections on transition metal complexes, lanthanide and actinide elements, inorganic polymers, and organometallic compounds. We have used the Système International d'Unités (SI) throughout the text as is now the custom in scientific publications.

We hope that the book will be of interest to both students and teachers particularly those wishing to approach inorganic chemistry from a quantitative standpoint. References to original papers and textbooks are listed at the ends of each chapter to enable students to make a deeper study of the subject, if desired.

We thank our colleagues and wives for their assistance and tolerance during the writing of this book.

S.J.A., Exeter
G.B., Wolverhampton

Acknowledgements

The authors are grateful to the following for permission to reproduce figures:

Figure 2.2. L. K. B. Instruments Limited
Figure 2.3. Dr. A. S. Carson and the Chemical Society
Figures 4.3, 4.4 and 4.5. Professor G. Wilkinson, FRS and John Wiley & Sons Inc.
Figures 5.7 and 5.8. Dr. M. L. H. Green and Methuen and Company Limited
Figure 6.2. Verlag Chemie GmbH.
Figure 6.6. Dr. B. R. Currell and the Chemical Society
Figure 7.4 and 7.5. Professor S. N. Vinogradov and Dr. R. H. Linnell and Van Nostrand Reinhold Company Limited

Chapter 1

Types of Bonding

1.1 Introduction

Chemistry is concerned with the interactions of atoms and molecules which comprise chemical substances. It is, therefore, of great interest to be able to make a quantitative study of these interactions. The most useful approach to such a study is through thermodynamics which is a science concerned, in part, with the measurement and calculation of the energy changes which occur during chemical reactions.

In order to make the text as self-contained as possible, we precede the discussion of thermodynamics by a brief summary of general bonding theory. A knowledge of atomic structure is assumed. Further discussions of bonding theory are to be found both in the text and in the recommended reading at the end of this chapter.

1.2 Ionic Bonding [1]

Purely ionic substances consist of cations and anions, although in the majority of cases, there is always some degree of covalent bonding. Cations can be formed most easily from those atoms which have only a weak attraction for their electrons, while anions form most readily from atoms which strongly attract electrons. Two quantities have been used to measure quantitatively these attractions – ionization energies and electron affinities respectively.

1.2.1 Ionization Energies (or Potentials)

An ionization energy may be defined as the energy required to remove completely an electron from a gaseous atom, molecule, or ion without imparting any kinetic energy to either species. Any particular species may have a series of ionization energies, referred to as the first, second, third, etc. All ionization energies of chemical interest are positive, and successive values increase since it becomes progressively more difficult to remove electrons from positive ions. We shall be concerned only with the first, second and, possibly, third ionization energies of gaseous *atoms*. A discussion of the thermodynamic aspects of ionization energies appears in Section 2.2.2.

Table 1.1. First Ionization Energies (I_1/kJ mol^{-1}) of some Representative Elements

IA	IIA	IIIB	IVB	VB	VIB	VIIB	VIIIB
H 1310							He 2372
Li 519	Be 900	B 799	C 1088	N 1406	O 1314	F 1682	Ne 2079
Na 498	Mg 736	Al 577	Si 787	P 1063	S 1000	Cl 1255	Ar 1519
K 418	Ca 590	Ga 577	Ge 782	As 967	Se 941	Br 1142	Kr 1351
Rb 402	Sr 548	In 556	Sn 707	Sb 833	Te 870	I 1008	Xe 1172
Cs 377	Ba 502	Tl 590	Pb 715	Bi 774	Po —	At —	Rn 1038

The first ionization energy (I_1) of atoms does not show a regular periodic variation, as may be seen from Table 1.1. This is because at least three factors contribute to the magnitude of I_1. These are:

(a) *Atomic radius:* This is exemplified in Groups IA and IIA in which there is a steady decrease in I_1 with increasing atomic mass. The outermost *ns* electrons are subjected to an attractive potential, V, of the form

$$V = -Z_{\mathbf{eff}} \, . \, e^2/(r \, . \, 4\pi\epsilon_0)$$

In this equation, ϵ_0 is the permittivity of free space, e is the electron charge, r is the mean distance of the *ns* electron from the nucleus (assumed equal to the atomic radius) and $Z_{\mathbf{eff}}$ is the effective nuclear charge. The latter varies only slightly on descending Groups IA and IIA since we are dealing with electrons which lie beyond *completed* quantum shells which shield outer electrons from the nucleus quite effectively. V and, therefore, I_1 decrease in the indicated order.

(b) *Incomplete electronic shielding:* All things being equal, we might expect the increase in nuclear charge, on moving along a period to be exactly compensated by an increase in electronic shielding. This is not so, as shown in Table 1.1. In the lithium to neon period there is a *general* increase in I_1 which indicates that electrons are imperfect in shielding the outermost electrons from the nucleus.

(c) *The type of electron removed*: The increase in I_1 is, however, not smooth and we observe the sequence of ionization energies:

$$\mathrm{Li} < \mathrm{Be} > \mathrm{B} < \mathrm{C} < \mathrm{N} > \mathrm{O} < \mathrm{F} < \mathrm{Ne}$$

The discontinuities are associated with the addition of the first and fourth 2*p*-electrons respectively. The radial part of the 2*s* orbital lies partly within that of the 2*p* orbital and is said to increase the 'screening' from the nucleus of the electron in the latter. The second discontinuity is associated with the removal of an electron from an orbital which already contains one electron. In this situation electron-electron repulsions are high.

The lowest values of I_1 are found for the heavier elements of Groups IA, IIA and IIIB and it is these which most easily ionize in chemical reactions to form cations.

1.2.2 Electron Affinities

During the formation of an ionic compound from the elements, it is frequently the case that the whole or part of the energy required to form the cation is provided by the exothermic formation of the anion. The enthalpy change for the reaction in which an electron is added to a gaseous atom, molecule, or ion is known as the *electron affinity* of the species. Electron affinities, A, as defined more fully in Section 2.2.2, are usually negative quantities since most of the common ions can readily accept electrons owing to incomplete shielding of the nucleus. Positive values of A do exist, however, particularly for metals and negatively charged species. Table 1.2 lists some typical values of A for atoms and ions. Note that some texts use an electron affinity defined as the negative of our A.

From the discussion in Section 1.2.1 we expect A to be more negative for the lighter elements of Groups VB, VIB, and VIIB; this is seen to be true. It is difficult to enter into a detailed discussion of the variation in A with atomic number since electron affinities, unlike ionization energies, are not easily

Table 1.2. Electron Affinities (A/kJ mol^{-1}) of some Gaseous Atoms

F	−350	Li	−52
Cl	−365	Na	−21
Br	−342	He	+54
I	−313	Ne	+99
O	−142	O^-	+844
S	−200	S^-	+532
H	−72		
C	−121		
N	+31		

subjected to direct experimental measurement and may, therefore, be approximate in many cases.

1.2.3 Structures of Simple Ionic Substances

Crystals of ionic compounds are often highly symmetric, and this is a reflection of the highly symmetric arrangement of cations and anions which constitute such compounds.

A common feature of many ionic inorganic compounds is that the cations or anions are *close-packed.* A close-packed layer of spheres (to represent spherical ions) is shown in Fig. 1.1(a). This arrangement is the most efficient

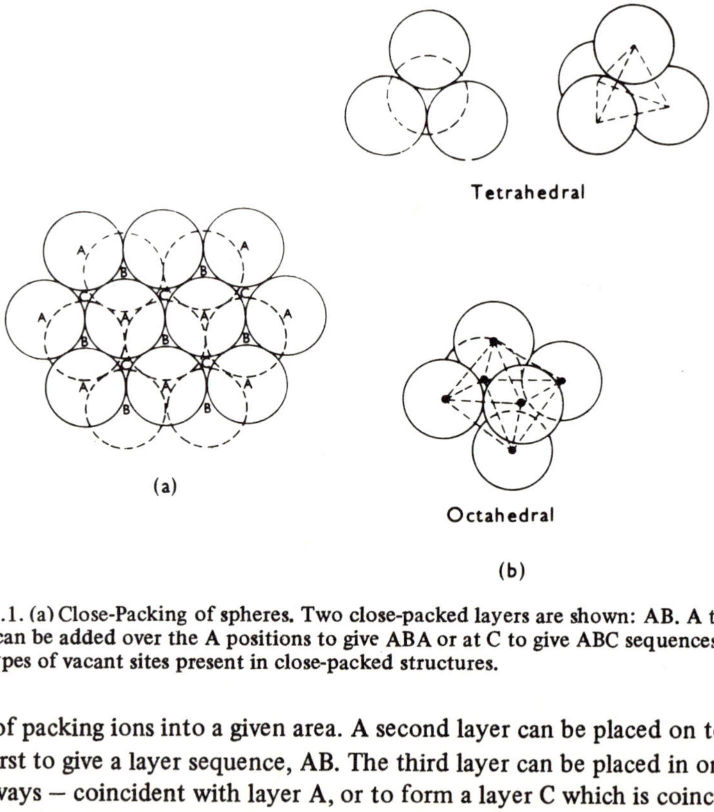

FIG. 1.1. (a) Close-Packing of spheres. Two close-packed layers are shown: AB. A third layer can be added over the A positions to give ABA or at C to give ABC sequences. (b) Types of vacant sites present in close-packed structures.

way of packing ions into a given area. A second layer can be placed on top of the first to give a layer sequence, AB. The third layer can be placed in one of two ways – coincident with layer A, or to form a layer C which is coincident with neither A nor B. In each of these stacking sequences the layers are placed such that the minimum volume is occupied. These arrangements of the close-packed layers – ABABAB etc., and ABCABCABC etc. – are called *hexagonal close-packing* (hcp) and *cubic close-packing* (ccp), respectively.

Either type of arrangement still leaves cavities between the ions. There are two types of cavity, and these are illustrated in Fig. 1.1(b). The first type is surrounded octahedrally by six spheres and is called an '*octahedral hole*'. The second type is called a '*tetrahedral hole*' and is surrounded tetrahedrally by four spheres. There are N octahedral holes and $2N$ tetrahedral holes for N close-packed spheres or ions. Some examples of ionic substances are listed below. The close-packed ion is underlined and the type of hole (oct. =

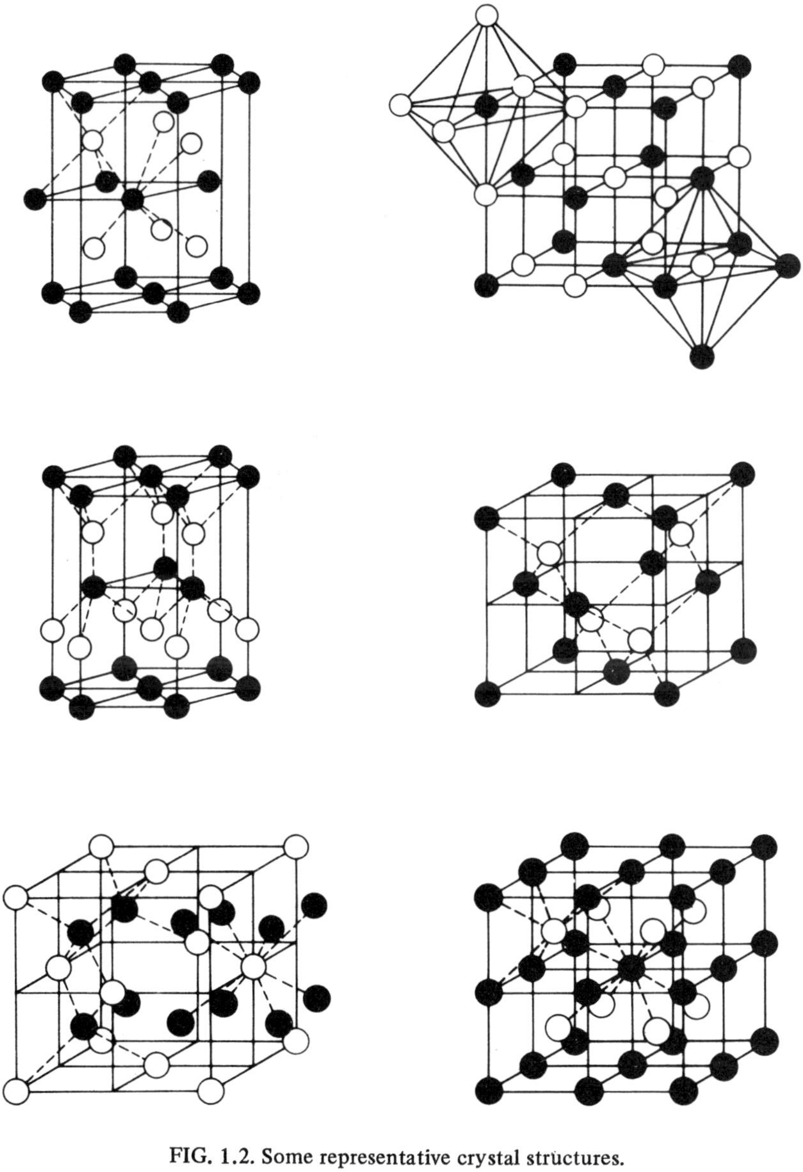

FIG. 1.2. Some representative crystal structures.

octahedral, tet. = tetrahedral), occupied by the other ion is indicated. For example, NaCl has a rock-salt structure; the Cl^- ions are cubic close-packed with the Na^+ ions in all of the octrahedral holes.

Structure	*Formula*	*Remarks*	*Examples*
Rock Salt	A^+B^- (ccp)	A^+ in oct. holes	NaCl
Zinc Blende	$A^{2+}B^{2-}$ (ccp)	A^{2+} in $\frac{1}{2}$ tet. holes	ZnS (low temp.)
Fluorite	$A^{2+}B_2^-$ (ccp)	B^- in tet. holes	CaF_2
Wurtzite	$A^{2+}B^{2-}$ (hcp)	A^{2+} in $\frac{1}{2}$ tet. holes	ZnS (high temp.)
Cadmium Iodide	$A^{2+}B_2^-$ (hcp)	A^{2+} in $\frac{1}{2}$ oct. holes	CdI_2
Cadmium Chloride	$A^{2+}B_2^-$ (ccp)	A^{2+} in $\frac{1}{2}$ oct. holes	$CdCl_2$

This is not a complete list and there are some structures which are not based on close-packed arrangements of ions. Some examples of both types are shown in Fig. 1.2 and, for a fuller discussion, the reader is referred to one of the texts listed at the end of this chapter [1].

The symmetrical arrangements of oppositely-charged ions is responsible for the high thermal stability of many ionic compounds. A more complete discussion of the energetics of ionic bonding is included in Chapter 3.

1.3 Covalent Bonding [2, 3]

When the conditions for ionic bonding, described in Section 1.2, are not satisfied, covalent bonding results if the atomic orbitals of the combining atoms can overlap sufficiently well. The wave-mechanical treatment of atoms yields a wave function, ϕ, which completely describes the behaviour of an electron in an atom. In an atom, ϕ is expressed in terms of functions which are dependent on the three spatial quantum numbers (n, l, and m) and the spin quantum number (s). The probability of finding an electron in a volume of space, dv, is given by $\phi^2 dv$. This results in the 'normalization condition',

$$\int \phi^2 \, dv = 1 \quad (1.1)$$

in which the integral is over the whole of space.

A bond is formed by the overlap of adjacent atomic orbitals. A useful approximate form of the resultant *molecular orbital* is provided by a method known as the 'Linear Combination of Atomic Orbitals', abbreviated to LCAO. Within this approximation, the form of the molecular orbital, ψ, is given by equation (1.2) for a diatomic molecule, where ϕ_A and ϕ_B respresent the atomic orbitals

$$\psi = N^{-1/2}(c_A\phi_A + c_B\phi_B) \quad (1.2)$$

centred on atoms A and B, and c_A, c_B are the numeric coefficients of each orbital. N is a normalization constant introduced to satisfy an equation similar to (1.1) and is given by

$$N = c_A^2 + c_B^2 + 2c_A c_B \int \phi_A \phi_B \mathrm{d}v \tag{1.3}$$

The last term is often neglected since usually

$$(c_A^2 + c_B^2) > 2c_A c_B \int \phi_A \phi_B \mathrm{d}v$$

If we apply (1.2) to a homonuclear molecule, then ϕ_A and ϕ_B contribute equally to ψ, and we obtain (since both positive and negative combinations are permitted)

$$\psi_1 = \frac{1}{\sqrt{2}}(\phi_A + \phi_B); \quad \psi_2 = \frac{1}{\sqrt{2}}(\phi_A - \phi_B) \tag{1.4}$$

The molecular orbital, ψ_1, is more stable than ψ_2 as can be derived from the wave equation when written in the Hamiltonian form (1.5), where E is the orbital energy

$$\mathbf{H}\psi = E\psi \tag{1.5}$$

and **H** is the Hamiltonian operator which, for a hydrogen atom, takes the following simple form:

$$\mathbf{H} = -\frac{h^2}{8\pi^2 m_e}\left(\frac{\partial^2}{\partial x^2} + \frac{\partial^2}{\partial y^2} + \frac{\partial^2}{\partial z^2}\right) - \frac{e^2}{r}$$

h is Planck's constant and m_e is the mass of the electron of charge e at a distance r from the nucleus; x, y, z, are the cartesian coordinates of the electron, referred to the nucleus.

For a many-electron atom **H** is more complex, owing to inter-electron repulsive terms. Applying (1.5) to (1.2) we obtain (1.6), in which the exact form of **H** is not defined:

$$\mathbf{H}(c_A\phi_A + c_B\phi_B) = E(c_A\phi_A + c_B\phi_B) \tag{1.6}$$

Let us multiply (1.6) firstly by ϕ_A and secondly by ϕ_B and integrate over the whole of space using the conditions and notations:

$\int \phi_A^2 \mathrm{d}v = \int \phi_B^2 \mathrm{d}v = 1$ (normalization)

$S = \int \phi_A \phi_B \mathrm{d}v$ (overlap integral)

$\beta = \int \phi_A \mathbf{H} \phi_B \mathrm{d}v = \int \phi_B \mathbf{H} \phi_A \mathrm{d}v$ (resonance integral)

$\alpha_A = \int \phi_A \mathbf{H} \phi_A \mathrm{d}v$
$\alpha_B = \int \phi_B \mathbf{H} \phi_B \mathrm{d}v$ } (Coulomb integrals, equal to α for a homonuclear molecule)

The results of this parameterization are, for a homonuclear molecule,

$$\begin{aligned} c_A(\alpha - E) + c_B(\beta - E\,.\,S) &= 0 \\ c_A(\beta - E\,.\,S) + c_B(\alpha - E) &= 0 \end{aligned} \qquad (1.7)$$

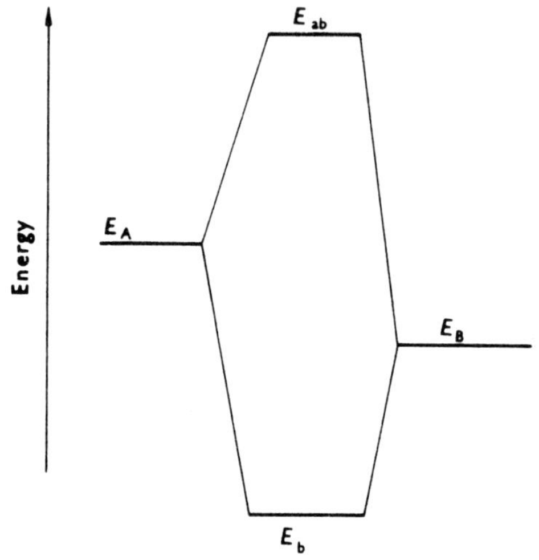

(a)

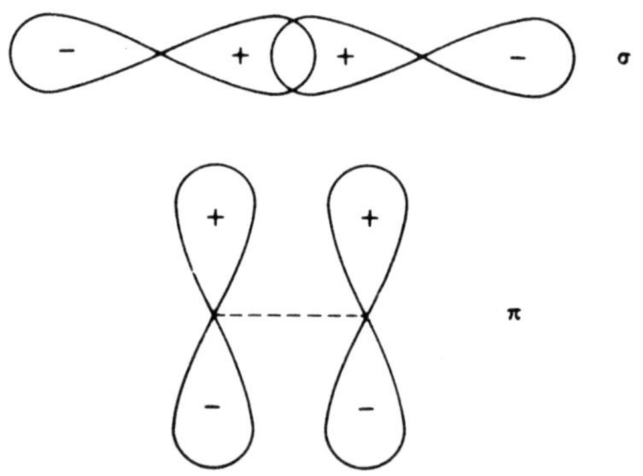

(b)

FIG. 1.3. (a) Bonding and antibonding levels arising from covalent bonding (b) σ and π overlap of p-atomic orbitals.

From the equations (1.7), we obtain

$$E_b = (\alpha + \beta)/(1 + S) \tag{1.8a}$$

$$E_{ab} = (\alpha - \beta)/(1 - S) \tag{1.8b}$$

$$\psi_b = \frac{1}{\sqrt{2}}(\phi_A + \phi_B) \tag{1.9a}$$

$$\psi_{ab} = \frac{1}{\sqrt{2}}(\phi_A - \phi_B) \tag{1.9b}$$

ψ_b and ψ_{ab} are the 'bonding' and 'antibonding' orbitals respectively. Note that β is inherently a negative quantity and, also, from equations (1.8) that $(E_{ab} - \alpha)$ is greater than $(\alpha - E_b)$. In Fig. 1.3 we summarize the results of this discussion and also illustrate the 'end-on' and 'sideways' overlap of atomic orbitals to give σ and π molecular orbitals.

The strength of a bond is expressed quantitatively by (1.8a) from which we can see that the lowering of E_b below E_A is governed primarily by β.

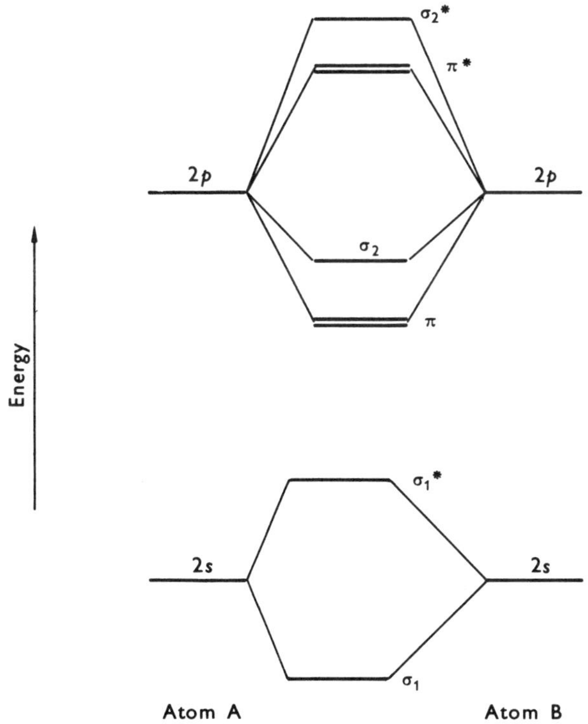

FIG. 1.4. An approximate molecular orbital energy level diagram for a diatomic first period molecule. (Note that σ_2 and π are reversed for O_2, F_2 and the hypothetical Ne_2.)

This is roughly proportional to S, the overlap integral, and leads to the 'criterion of maximum overlap'. This states that the strength of a bond formed between two atoms is proportional to the overlap between atomic orbitals, expressed quantitatively by the overlap integral.

Of the two types, π-bonding is generally weaker than σ-bonding for a given bond length and pair of orbitals since, in the former, the combining orbitals are constrained to be further apart.

A useful diagram is shown in Fig. 1.4 which depicts the molecular orbital energy levels for the homonuclear diatomic molecules from H_2 to N_2. Beyond N_2 the energies of the σ_2 and π levels are reversed.

The orbitals are labelled σ and π to denote their symmetries; levels with an asterisk, such as π^*, are antibonding. The diagram needs to be amended for heteronuclear molecules since the energies of the combining orbitals are not necessarily identical.

1.4 Hydrogen Bonding [4]

This is a specialized form of the more general *dipole–dipole* bonding in which bonds are formed between atoms of opposite polarity. Such bonding

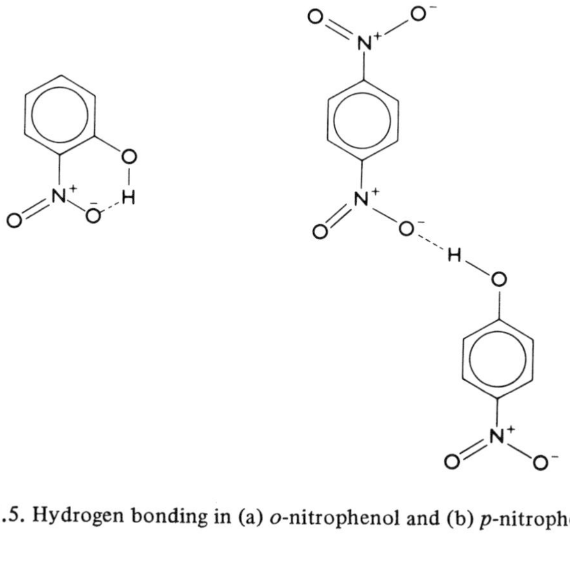

FIG. 1.5. Hydrogen bonding in (a) *o*-nitrophenol and (b) *p*-nitrophenol.

is primarily electrostatic. Hydrogen bonding is particularly important since hydrogen is often covalently bonded to electronegative atoms, resulting in a considerable induced positive charge at the hydrogen atom. This *positive* charge is attracted to any adjacent atom which bears a *negative* charge. For example, in *o*-nitrophenol, the hydroxylic proton is said to be *intramolecu-*

larly hydrogen bonded to an oxygen atom of a nitro group *in the same molecule.* In *p*-nitrophenol, the hydroxylic proton is *intermolecularly* hydrogen bonded to a nitro group of a different molecule. This is shown in Fig. 1.5.

Hydrogen bonding is often manifested in the physical or chemical properties of a compound. For example, the enthalpy of vaporization and boiling temperature of water are higher than the values predicted by extrapolation through the series H_2Te, H_2Se, H_2S. From these results it is inferred that molecules in liquid water are associated owing to H . . . O hydrogen bonding of the intermolecular type. Association can be more complete than this as, for example, in the dimerization of acetic acid in the gas phase or in the formation of the solid 1:1 molecular complex between urea and succinimide. Hydrogen bonding is also responsible for the polymeric structure of boric acid in the solid or molten phases. A more detailed discussion follows in Chapter 7.

1.5 Solvation of Ions

Solvent molecules are often strongly bonded to positively charged metal ions. An electrostatic model, due to Born, of this solvation behaviour has been widely used. Although there are certain theoretical objections to this simple model, it still has some value – especially in the prediction of trends in behaviour.

The Gibbs free energy change on transferring N charged spheres into a medium of dielectric constant, ϵ, is given by

$$\Delta G = \frac{-Ne^2}{r \, . \, 4\pi\epsilon_0}\left(1 - \frac{1}{\epsilon}\right)$$

where ϵ_0 is the permittivity of free space and r is the radius of a sphere of charge, e. ΔG is increased for small ions of high charge in a solvent of high dielectric constant. The application of this model will be discussed in Chapter 7.

When we discuss non-aqueous solvents in Chapter 7 it will become clear that the donor strength of a solvent is an important factor in determining the behaviour of solutes. The donor strength of a solvent can be estimated thermochemically from the enthalphy change for the interaction of the solvent with an electron acceptor such as $SbCl_5$ in high dilution in a non-interacting solvent. Measurements of the enthalpy change of reaction (1.10)

$$B\!: + SbCl_5 = B\!: SbCl_5 \qquad (1.10)$$

have yielded the following order of solvent donor strengths:

pyridine > $(CH_3)_2SO$ > *N, N*-dimethylformamide > tetrahydrofuran > water > acetone > propionitrile > acetonitrile > nitrobenzene

Strong donor solvents can cause the fission of covalent bonds in un-ionized solvents to form ions:

$$B: + M{-}X \rightarrow [B{-}M{-}X] \rightarrow B{-}M^{n+} + X^{n-}$$

Therefore, the enthalpy of solvation can provide the energy for bond fission in some cases and this occurs increasingly with (a) decreasing M–X bond strengths (b) increasing polarizability of the M–X bond and (c) increasing solvent donor strength.

1.6 Soft and Hard Acids and Bases [5]

In Section 1.5, the impression may have been given that solvation is a primarily electrostatic process. While this may be true in some cases, there is ample evidence to show that covalent bonding is often important. Furthermore, the coordination of donor molecules (*ligands*) to acceptor atoms, ions or molecules is more complicated than might at first be thought. A useful rationale is provided by the 'Principle of Soft and Hard Acids and Bases', abbreviated to SHAB. In order to understand this principle, consider the equilibrium for the exchange reaction:

$$CH_3Hg(H_2O)^+(aq) + BH^+(aq) \rightleftharpoons CH_3HgB^+(aq) + H_3O^+(aq) \qquad (1.11)$$

A large equilibrium constant for this reaction will be taken to mean that the base, B, prefers to coordinate to the electron-acceptor (Lewis Acid) $CH_3Hg^+(aq)$. Conversely, a small equilibrium constant means that the protonated base is more stable and that B prefers to coordinate to the proton.

There is a wealth of data on equilibria involving the $CH_3Hg^+(aq)$ ion and the equilibrium constants for reaction (1.11) are in the order

$$CH_3 > I^- > Br^- > Cl^- > S_2O_3^{2-} > S^{2-} > (C_2H_5)_3P > CN^- > F^- > HPO_3^{2-} \sim NH_3 > OH^-$$

Clearly the donor atoms N, F, and O prefer to coordinate to $H^+(aq)$ whereas the donor atoms C, I, Br, Cl, S, and P prefer to donate their electron pairs to $CH_3Hg^+(aq)$. Bases containing the first group of atoms are termed 'hard' and those containing the latter group are termed 'soft'. If we study similar equilibria for a range of metal atoms and ions (Lewis acids) a similar classification results: some metals form stronger bonds with soft bases such as CH_3Hg^+ whilst others form strong bonds with hard bases. The former are termed 'soft acids' and the latter, 'hard acids'. Examples of the two types are listed below:

Hard acids: H^+, Li^+, Na^+, K^+, Be^{2+}, Mg^{2+}, Ca^{2+}, Al^{3+}, Sc^{3+}, Si^{4+}

Soft acids: Cu^+, Ag^+, Au^+, Tl^+, Pd^{2+}, Pt^{2+}, I^+, I_2, M^0 (metal atoms and bulk metals)

Hard bases: H_2O, OH^-, F^-, $CH_3CO_2^-$, PO_4^{3-}, RO^-
Soft bases: R_2S, RS^-, R_3P, C_2H_4, H^-, R^-

Hard acids and bases usually contain small acceptor and donor atoms, often with large charges that are difficult to polarize. Soft acids are more polarizable, usually involving larger atoms with smaller charges. We recognize a hard-acid/hard-base combination as that which leads to a large percentage of ionic or electrostatic bonding; soft-acid/soft-base combinations are those usually associated with covalent bonding. A very simple rule has been enunciated by R. G. Pearson [5] concerning the stability of acid-base complexes: 'hard acids prefer to bind to hard bases; soft acids prefer to bind to soft bases'. This is a semi-quantitative rule and expresses an intuitive feeling that the atomic conditions for *either* electrostatic *or* covalent bonding must be satisfied by *both* bonding partners.

Although there are exceptions, a surprising number of quite diverse chemical reactions have been rationalized in terms of this simple rule. Exceptions may be expected and do, in fact, occur when solvents other than water are considered. A large amount of experimental data on aqueous systems has been correlated by Pearson to produce a quantitative scale of softness. Values of a parameter, σ, are given in Table 1.3 for a range of ions, the value of σ *decreasing* with increasing softness.

Table 1.3. Values of Pearson's Softness Parameter. (Decreasing σ indicates increasing softness for aqueous systems)

M^+	σ	M^{2+}	σ	M^{3+}	σ	M^{4+}	σ
Li	0·247	Be	0·172	Al	0·136	Ti	0·102
Na	0·211	Mg	0·167	Sc	0·140	Zr	0·121
K	0·232	Ca	0·180	Y	0·147		
Rb	0·229	Sr	0·172	La	0·171		
Cs	0·218	Ba	0·184	Cr	0·107		
H	0·149	Cr	0·142	Fe	0·097		
Cu	0·112	Mn	0·124	Ga	0·099		
Ag	0·073	Fe	0·127	In	0·100		
Au	0·044	Co	0·130				
Ga	0·267	Ni	0·126				
In	0·213	Cu	0·104				
Tl	0·215	Zn	0·115				
		Cd	0·081				
		Hg	0·064				
		Sn	0·148				
		Pb	0·131				

REFERENCES

Crystal Structures

[1] Cotton, F. A., and Wilkinson, G., *Advanced Inorganic Chemistry*, 2nd edn., Interscience, New York (1966).

Covalent Bonding

[2] Cartmell, E., and Fowles, G. W. A., *Valency and Molecular Structure*, 2nd edn., Butterworths, London (1961).

[3] Gray, H. B., *Electrons and Chemical Bonding*, Benjamin, New York (1964).

Hydrogen Bonding

[4] Vinogradov, S. N., and Linnell, R. H., *Hydrogen Bonding*, Van Nostrand Reinhold, New York (1971).

Soft and Hard Acids and Bases

[5] Pearson, R. G., *J. Chem. Educ.*, **45**, 581, (1968) and also *ibid. idem*, **45**, 643, (1968).

Chapter 2

Thermodynamics and Experimental Methods

2.1 Chemical Reactions

Reactions such as

$$cC + dD \ldots = pP + qQ \ldots$$

may be written in a convenient general form

$$0 = \Sigma_B \nu_B B$$

where B represents any reacting species and ν_B is the stoichiometric coefficient of B in the equation, positive for products, negative for reactants. Molar changes in thermodynamic functions such as internal energy ΔU, enthalpy ΔH, entropy ΔS, Gibbs function (or free energy) ΔG, and heat capacity C_p and C_v may then be related to the molar thermodynamic properties of the reacting species as follows:

$$\Delta U = \Sigma_B \nu_B U_B: \quad \Delta H = \Sigma_B \nu_B H_B: \quad \Delta S = \Sigma_B \nu_B S_B: \quad \Delta G = \Sigma_B \nu_B G_B$$

$$\Delta C = \Sigma_B \nu_B C_B$$

Conventions relating to standard states are discussed in Section 2.2.1.

The extent to which a reaction has progressed when equilibrium is attained is measured by the equilibrium constant K which, for a gaseous reaction, is related to the fugacities f_B of the reacting species by

$$K_f = \Pi_B (f_B)^{\nu_B}$$

For reactions which occur in the solid or liquid phase the equilibrium constant is more conveniently related to the activities a_B of the reacting species

$$K_a = \Pi_B (a_B)^{\nu_B}$$

It is usual to refer to the equilibrium constants for the formation of complexes in solution as stability constants. When several complexes coexist, such as those formed by the stepwise addition of ligand molecules to a central metal ion, it is important to note whether the stability constant refers to the overall formation from free metal ion and ligand or to a particular step. Overall constants are sometimes given the symbol β. An alternative measure of the position of equilibrium is the molar standard Gibbs function change $\Delta G^{\ominus}$ since this is related to the equilibrium constant by

$$\Delta G^{\ominus} = -RT \ln K$$

2.2 Thermodynamic Quantities and Conventions

2.2.1 Standard States of Tabulated Data

Tables of chemical thermodynamic data [1–4], extracts from which appear throughout this text, usually list values for substances at 298·15 K in their standard states (indicated by the superscript ^{o} or $^{\ominus}$ on the thermodynamic symbol). For a pure liquid or solid the standard state refers to the substance under a pressure of 101·325 kPa (= 1 atm). For a gas the standard state is the hypothetical ideal gas state at a fugacity of 101·325 kPa, in which state the enthalpy is that of the real gas at the same temperature and zero pressure. The abbreviations (s), (c), (l), (g) following the formula for a substance are used to distinguish the solid or crystalline, liquid, and gaseous states, respectively.

Standard thermodynamic functions of formation $\Delta_f G^{\ominus}$, $\Delta_f H^{\ominus}$, and $\Delta_f S^{\ominus}$ refer to the change in the appropriate thermodynamic quantity when one mole of the substance in its standard state is formed from the elements each in its standard reference state, which is almost always the thermodynamically stable form of the element in its standard state at 298·15 K. A further convention is that enthalpies of formation of gaseous ionic species are usually calculated with $\Delta_f H^{\ominus} = 0$ for the gaseous electron.

The standard state for a solute in aqueous solution is usually a hypothetical ideal solution of unit molality, and values of thermodynamic properties of individual ions in aqueous solution are related by convention to the zero values of $\Delta_f H^{\ominus}$, $\Delta_f G^{\ominus}$, $S^{\ominus}$, and $C_p^{\ominus}$ for the hydrogen ion in its aqueous standard state.

Frequently only $\Delta_f G^{\ominus}$ and $\Delta_f H^{\ominus}$ are tabulated, and the additional quantity $S^{\ominus}$, the 'thermal' entropy, is given. This is the entropy of the substance in its standard state relative to the entropy of a perfect crystal of the substance at 0 K, which is zero by convention.

2.2.2 Changes in Thermodynamic Functions

Values of $\Delta G^{\ominus}$, $\Delta H^{\ominus}$, and $\Delta S^{\ominus}$ for a reaction are related to the tabulated data for the reacting species by

$$\Delta G^{\ominus} = \Sigma_B \nu_B \Delta_f G^{\ominus}(B): \quad \Delta H^{\ominus} = \Sigma_B \nu_B \Delta_f H^{\ominus}(B): \quad \Delta S^{\ominus} = \Sigma_B \nu_B S^{\ominus}(B)$$

A number of thermodynamic quantities appearing throughout this text are given special names. Table 2.1 summarizes these.

Table 2.1. Thermodynamic Quantities with Special Names (A simple compound MA_n is used here for illustration)

Quantity	*Definition*	*Method of measurement*
Gibbs function of reaction	$\Delta G^{\ominus} = \Sigma_B \nu_B \Delta_f G^{\ominus}(B)$	equilibrium constants: $\Delta G^{\ominus} = -RT \ln K$ or $\Delta G = \Delta H - T\Delta S$ (2.5.1)
enthalpy of reaction	$\Delta H^{\ominus} = \Sigma_B \nu_B \Delta_f H^{\ominus}(B)$	calorimetry; temperature dependence of K: $\frac{d \ln K}{d(1/T)} = -\frac{\Delta H^{\ominus}}{R}$ (2.5.1–2.5.4)
entropy of reaction	$\Delta S^{\ominus} = \Sigma_B \nu_B S^{\ominus}(B)$	calorimetric measurement of C_p: $dS = C_p \, d \ln T$
heat capacity change, const. p const. V	$\Delta C_p = \Sigma_B \nu_B C_p(B)$ $\Delta C_v = \Sigma_B \nu_B C_v(B)$	calorimetry
enthalpy of fusion of B	$\Delta_s^l H^{\ominus} = \Delta H^{\ominus}$ for $B(s) = B(l)$	calorimetry; temperature dependence of vapour pressure using Clausius–Clapeyron equation: $\frac{d \ln p}{d(1/T)} = -\frac{\Delta H^{\ominus}}{R}$
enthalpy of vaporization of B	$\Delta_l^g H^{\ominus} = \Delta H^{\ominus}$ for $B(l) = B(g)$	
enthalpy of sublimation of B	$\Delta_s^g H^{\ominus} = \Delta H^{\ominus}$ for $B(s) = B(g)$	

Quantity	*Definition*	*Method of measurement*
lattice energy of $MA_n(s)$	$U = \Delta U^{\ominus}(0\ K)$ for $MA_n(s) = M^{n+}(g) + nA^-(g)$	calculation from theoretical equation or from cycle. Often assumed $U(0\ K) \approx \Delta H^{\ominus}(298{\cdot}15\ K)$ (2.2.2)
bond dissociation energy of M–A bond in MA_n	$D(M{-}A) = \Delta H^{\ominus}$ for $MA_n(g) = MA_{n-1}(g) + A(g)$	gas phase kinetics; mass spectra; indirect calculation

Table 2.1. (*continued*).

bond energy term of M–A bond in $MA_n^{(a)}$	E(M–A) = $\Delta H^{\ominus}/n$ for $MA_n(g) = M(g) + nA(g)$	indirect calculation from cycle
coordinate bond dissociation energy of M–A bond in MA_n	CBE(M–A) = $\Delta H^{\ominus}$ for $MA_n(g) = MA^+_{n-1}(g) + A^-(g)$ or, if A is a neutral ligand $MA^+_n(g) = MA^+_{n-1}(g) + A(g)$	indirect calculation from cycle
mean coordinate bond dissociation energy of M–A bond in MA_n	$\overline{CBE}$(M–A) = $\Delta H^{\ominus}/n$ for $MA_n(g) = M^{n+}(g) + nA^-(g)$	indirect calculation from cycle
ionization energy (or potential) of B	$I = \Delta U^{\ominus}(0\ K)$ for $B(g) = B^+(g) + e^-(g)$	spectroscopy; mass spectra. Often assumed $I \approx \Delta H^{\ominus}(298{\cdot}15\ K)$ (2.2.2) (Table 1.1)
electron affinity of B	$A = \Delta U^{\ominus}(0\ K)$ for $B(g) + e^-(g) = B^-(g)$	indirect calculation. Often assumed $A \approx \Delta H^{\ominus}$ (298·15 K) (2.2.2) (Table 1.2)
hydration enthalpy of gaseous ion $M^+(g)$	$\Delta_g^{aq}H^{\ominus}(M^-, g) = \Delta H^{\ominus}$ for $M^+(g) = M^+(aq)$	calorimetry and cycles
enthalpy of solution of solid compound MA_n	$\Delta_s^{aq}H^{\ominus}(MA_n, s) = \Delta H^{\ominus}$ for $MA_n(s) = MA_n(aq)$	calorimetry

NOTE (*a*): E(M–A) = mean bond dissociation energy $\bar{D}$(M–A) for a compound with only one type of bond.

The ionization energy (or potential) I and the electron affinity A defined in Table 2.1 are for one-electron changes. Equivalent expressions for changes of more than one electron can be written. Both I and A, and also lattice energy U refer to changes in internal energy at 0 K. They are related to the values of ΔU at 298·15 K by the Kirchoff equation

$$\Delta U(298{\cdot}15\ \mathrm{K}) - \Delta U(0\ \mathrm{K}) = \int_{0\ \mathrm{K}}^{298{\cdot}15\ \mathrm{K}} C_v\, \mathrm{d}T$$

in which, for I and A, it is usually sufficiently accurate to substitute the ideal monatomic gas value of $3/2R$ for C_v for each reacting species. Thus for an nth ionization energy

$$\Delta U(298{\cdot}15\ \mathrm{K})/\mathrm{kJ\ mol^{-1}} = \Sigma_n I + \frac{3n}{2}RT = \Sigma_n I + 0{\cdot}37n$$

For lattice energies this calculation can only be done if the temperature variation of the heat capacity of the solid is known. In the absence of such data, it is often assumed that

$$U(298{\cdot}15\ \mathrm{K}) \approx U(0\ \mathrm{K})$$

In general, the enthalpy change corresponding to a value of ΔU for a reaction is calculated from

$$\Delta H - \Delta U = \Sigma_\mathrm{B}\nu_\mathrm{B}(pV)_\mathrm{B} \approx \Sigma_\mathrm{B}\nu_\mathrm{B}^\mathrm{g}(RT)$$

usually by making the approximation shown in the second expression in which only gaseous species are significant. For an nth ionization energy, for example,

$$\Delta H(298{\cdot}15\ \mathrm{K})/\mathrm{kJ\ mol^{-1}} = \Sigma_n I + \frac{3n}{2}RT + nRT = \Sigma_n I + 0{\cdot}62n$$

Another frequent approximation, justifiable in view of the uncertainties often associated with experimental measurements of I, A, and U, is to assume that these quantities can be equated with the enthalpy change for the appropriate process at 298·15 K.

2.2.3 Simplification of Calculations

A useful simplification results from a combination of I or A at 298·15 K with enthalpy of sublimation or vaporization data to give enthalpies of formation of gaseous ions. For example

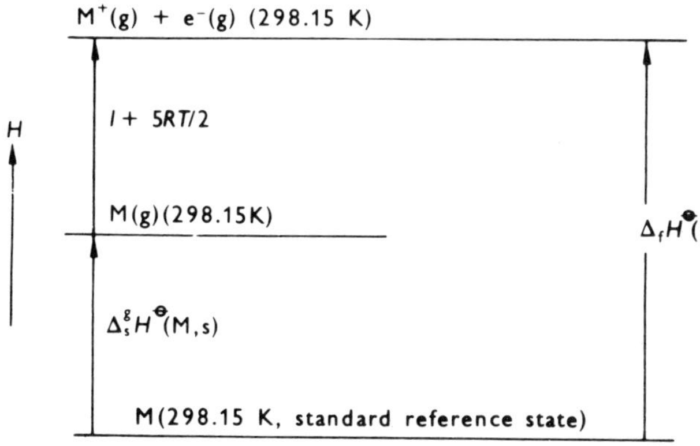

Enthalpies of formation of gaseous monatomic ions are to be found in the compilations listed at the end of this chapter. Table 2.2 gives equivalent data for some common polyatomic ions. Note also that Tables 1.1 and 1.2 list values of I and A, respectively, for the more common species, and Table 5.2 lists data for gaseous metals and organic radicals.

Table 2.2. Standard Enthalpies of Formation of Gaseous Ions at 298·15 K (Mainly from [3]).

Ion	$\Delta_f H^\ominus$/kJ mol^{-1}	Ion	$\Delta_f H^\ominus$/kJ mol^{-1}
ClO_3^-	−242	NO_2^-	−137
ClO_4^-	−383	NO_3^-	−335
BrO_3^-	−186	CO_3^-	−197
IO_3^-	−338	$HCOO^-$	−450
OH^-	−141	CH_3COO^-	−510
SO_4^{2-}	−745	CN^-	+53·6
CrO_4^{2-}	−761	CNS^-	−48·5
NH_2^-	+174	N_3^-	+142

2.3 Stability

2.3.1 Thermodynamic Stability [5]

The thermodynamic stability of a compound towards a particular decomposition is known if the value of the change in the standard Gibbs function, $\Delta G^\ominus$, or the equilibrium constant K ($\Delta G^\ominus = -RT \ln K$), is available for the reaction. A large positive value of $\Delta G^\ominus$ indicates stability of reactants while a large negative value indicates stability of products. An assessment of a compound's stability requires, therefore, a knowledge of $\Delta G^\ominus$ for all possible decomposition paths. The data necessary for this type of calculation are rarely available.

The change in Gibbs function is related to the enthalpy and entropy changes, for an isothermal process, by

$$\Delta G = \Delta H - T\Delta S$$

Although, for a particular reaction, all three quantities are sometimes available, it is common for only the enthalpy change to be known. Moreover, the enthalpy term expressed as a bond energy is the most appropriate quantity in discussions of bonding theories which attempt the calculation of bonding forces rather than thermodynamic stability.

2.3.2 Kinetic Stability

Thermodynamically unstable compounds do not necessarily undergo spontaneous decomposition. An immeasurably slow decomposition reaction would indicate kinetic rather than thermodynamic stability. The book by

Basolo and Johnson [6] gives an introduction to inorganic kinetics which will not be considered in the present text.

2.3.3 Interpretation of Thermodynamic Data

Thermodynamics does not itself give any insight into the nature of chemical bonding but it provides a means of re-expressing experimental results in ways that can be directly related to current bonding theories and models.

It is generally preferable when discussing the relation of theory to experiment to consider reactions in the gaseous phase where the energy changes are independent of the environmental effects of the liquid or solid phases. In the condensed phases, not only is the entropy term relatively important to the stability of the compound, but ΔH includes the energy of interaction of the species with the environment. In the gaseous phase the entropy changes are generally unimportant relative to the enthalpy changes and ΔH may be used as a measure of both bond energy and thermodynamic stability. For example, in the gaseous phase the total nickel-ammonia coordinate bond dissociation energy in the hexamminenickel (II) ion is $\Delta H \approx \Delta G$ for the reaction

$$[Ni(NH_3)_6]^{2+}(g) = Ni^{2+}(g) + 6NH_3(g)$$

$$CBE(Ni{-}NH_3) = 1770 \text{ kJ mol}^{-1}$$

In aqueous solution the values of ΔH and ΔG for the equivalent reaction

$$[Ni(NH_3)_6]^{2+}(aq) = Ni^{2+}(aq) + 6NH_3(aq)$$

are 102·9 kJ mol^{-1} and 49·8 kJ mol^{-1} respectively. Two important points in this second case are

(1) that the entropy term $T\Delta S$ = 53·1 kJ mol^{-1} is large relative to ΔH and so ΔG and ΔH differ considerably from each other;

(2) that the value of ΔH, to which we shall on occasion refer as a coordinate bond dissociation energy for an aqueous complex, actually represents the difference between the bond energy of the Ni–NH_3 bonds in the complex and the Ni–H_2O bonds in the aqueous nickel ion, better written as $[Ni^{II}(H_2O)_6]^{2+}$, both with additional water molecules more weakly bonded in outer layers.

In problems where the complications of the condensed phase become intractable, the thermochemical approach generally attempts to relate *trends* in thermodynamic quantities to aspects of bonding theory for a series of similar compounds. For species in aqueous solution, environmental effects can be reduced, or at least made constant, either by correcting measured values to relate to hypothetical reactions at infinite dilution or by making the measurements in a medium of constant ionic strength.

In any discussion of the strength of bonding in the gaseous phase, a further important point is that the energy of a bond in a gaseous compound may be related to ΔH for either of the two following processes

$$MA_n(g) = MA_{n-1}(g) + A(g) \quad \Delta H = D(M{-}A)$$

$$MA_n(g) = MA^+_{n-1}(g) + A^-(g) \quad \Delta H = CBE(M{-}A)$$

The high-temperature stability of inorganic compounds must concern processes of the first type since thermal dissociation into atoms is the more likely reaction. Likewise the successful bond energy schemes of organic chemistry relate to reactions of this type. On the other hand, some bonding models of inorganic chemistry, such as the hard sphere, polarizable sphere, and crystal field theories can most easily be related to processes of the second type. This does not imply that only ionic bonding can be discussed in this second way.

2.3.4 Spectroscopic and Kinetic Measurements of Energies

Spectroscopic measurements in the visible and ultraviolet regions are the source of data on ionization potentials and ligand-field splitting parameters (Chapter 4), both of which are regularly used in thermodynamic calculations. In addition, infra-red spectroscopic absorption frequencies can be roughly related to bond strengths, a strong bond having a higher vibrational frequency than a weak one. This is particularly useful for the comparison of a particular M–A bond in different environments, when thermodynamic measurements are not available. When bonds between different pairs of elements are to be compared, the *force constant*, determined spectroscopically, is the appropriate quantity to use.

Kinetic measurements yielding activation energies for bond breaking reactions may sometimes be related to ΔH for the same process if the activation energy for the back reaction is zero or can be estimated. This method has been widely applied to the study of the thermodynamics of organic free-radicals.

2.4 Thermodynamic Cycles

Only in a very small number of cases is it possible to measure directly the enthalpy or free energy changes associated with gas-phase reactions. Since this type of reaction is important for the development of bonding theories, it is necessary to relate gas-phase enthalpy and free energy changes to measurable quantities. The method of doing this is probably most easily seen on an energy diagram of the type illustrated in Fig. 2.1 for the calculation of the enthalpy of formation, $\Delta_f H^\ominus$(comp., g), (and hence the total coordinate bond dissociation energy) of the gaseous complex ion $[Co(NH_3)_6]^{2+}$ from gaseous Co^{2+} and NH_3. Calorimetric measurements supply the enthalpy of formation of the crystalline complex $[Co(NH_3)_6]Cl_2$ where the halide is present for

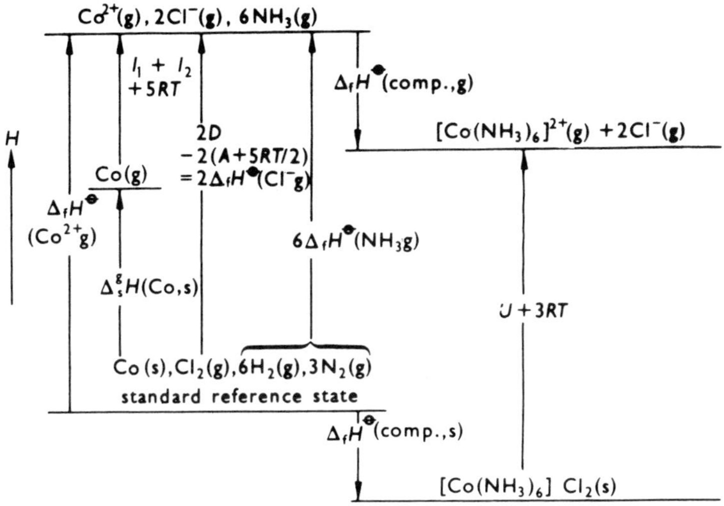

$$\Delta_s^g H(\text{Co, s}) + I_1 + I_2 + 5RT + 2D - 2(A + 5RT/2) + 6\Delta_f H^\ominus(\text{NH}_3, \text{g}) + \Delta_f H^\ominus(\text{comp., g}) - (U + 3RT) - \Delta_f H^\ominus(\text{comp., s}) = 0$$

or $$\Delta_f H^\ominus(\text{Co}^{2-}, \text{g}) + 2\Delta_f H^\ominus(\text{Cl}^-, \text{g}) + 6\Delta_f H^\ominus(\text{NH}_3, \text{g}) + \Delta_f H^\ominus(\text{comp., g}) - (U + 3RT) - \Delta_f H^\ominus(\text{comp., s}) = 0$$

FIG. 2.1. Enthalpy diagram for the calculation of gaseous-phase enthalpy changes. The lattice energy U is taken to be ΔU at 298·15 K. The relation between quantities is obtained from the relation for a complete cycle that $\Sigma\Delta H = 0$.

reasons of electrical neutrality. The other quantities required in the calculation are available from calorimetric and spectroscopic measurements with the exception of the lattice energy, U, which is obtained by calculation using a theoretical equation of the type discussed in Section 3.3.3. Since the enthalpy change depends only on the state of the reactants and products and not on the reaction path, a cyclic calculation of ΔH will result in $\Sigma\Delta H = 0$. This leads to the equation given beneath Fig. 2.1 from which $\Delta_f H^\ominus$ (comp., g) may be extracted.

The mean cobalt–ammonia coordinate bond dissociation energy is given by

$$\overline{CBE}(\text{Co–N}) = -\Delta_f H\,(\text{comp., g})/6$$

2.5 Experimental Methods

2.5.1 Equilibrium Constant Measurement

For the formation of complexes, particularly those in solution, a variety of analytical methods has enabled equilibrium constants (called stability constants) and hence Gibbs function changes to be determined [7]. The variation of these with temperature leads to values of the enthalpy changes, from

$$\frac{\mathrm{d}\ln K}{\mathrm{d}(1/T)} = -\frac{\Delta H^\ominus}{R}$$

but this method is generally considered to be inferior to direct calorimetric measurement of $\Delta H^{\ominus}$. Similar calculations on compounds and complexes in the solid phase are possible from measurements of dissociation pressures if decomposition into a gaseous product occurs. However, the majority of thermodynamic data on solid compounds is calorimetrically determined.

Further mention of the equilibria present in aqueous solution will be postponed until Section 4.3.

2.5.2 Reaction Calorimetry [8, 9]

If a suitable quantitative reaction can be found, the enthalpy of formation of a compound may be calculated from the calorimetrically measured enthalpy of reaction together with known thermochemical data on the other reacting species.

A typical reaction calorimeter is shown in Fig. 2.2(i) for the reaction of a

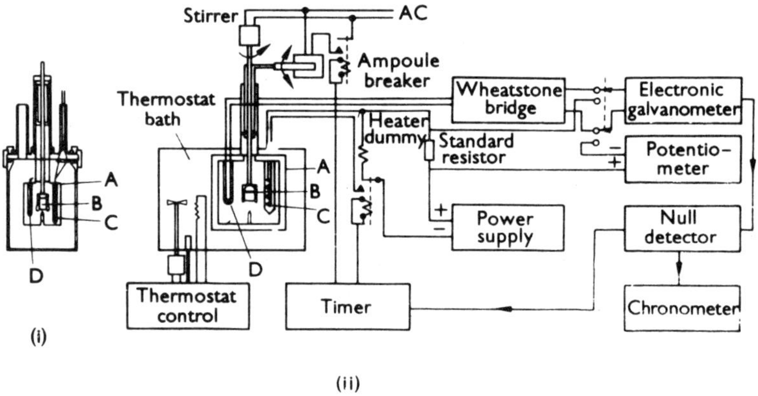

FIG. 2.2. Reaction calorimeter (i) and associated equipment (ii). A: reaction vessel, B: ampoule of reactant, C: calibration heater, D: thermistor for temperature measurement.

solid or liquid sample, contained in a glass ampoule, with a larger volume of liquid. Temperature measurement is by means of a resistance bridge, one arm of which is connected to a thermistor temperature probe contained in the reaction vessel. Precise thermostatting (± 0·005 K), electrical calibration and timing equipment, schematically shown in Fig. 2.2(ii), enables enthalpies of reaction to be determined to ± 0·1 per cent. Modifications of this type of calorimeter allow a continuous titration technique [10] to be used for the simultaneous determination of enthalpy and Gibbs function of stepwise formation of complexes in solution.

2.5.3 Bomb Calorimetry

Although bomb calorimetry, which involves burning a compound in a high-pressure atmosphere of oxygen, is widely used for determining the enthalpies of formation of organic compounds, it suffers from the serious limitation that inorganic compounds almost always produce either ill-defined oxides or mixtures of combustion products. Consequently the measured energy change (ΔU) cannot be ascribed to a particular reaction.

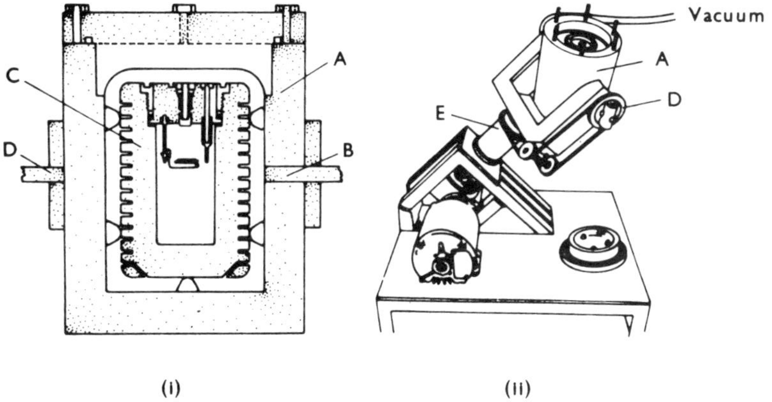

FIG. 2.3. Rotating bomb calorimeter (i) and equipment for rotation (ii). A: copper jacket, B: vacuum line and entry for electrical leads, C: copper calorimeter wound with resistance thermometer, D and E: axles for rotation. This aneroid type of calorimeter requires no thermostat; the jacket A maintains the constant temperature.

The development of rotating bomb calorimeters (Fig. 2.3) has greatly widened the scope of bomb calorimetry in the study of inorganic compounds and complexes. In addition to the compound and high-pressure oxygen, the calorimeter is charged with a few cubic centimetres of a suitable solution which will react with the combustion products to produce a final liquid of known composition. The whole calorimeter is rotated about two mutually perpendicular axes to ensure that the inner surface of the bomb is completely washed by liquid. This procedure effectively combines the functions of a combustion calorimeter and a solution calorimeter in one instrument. An example of this technique [11] is the combustion of germanium tetrabenzyl in a rotating bomb charged with oxygen and a little aqueous potassium hydroxide. The GeO_2 and CO_2 produced by combustion are dissolved in the KOH solution when rotation begins. In a comparison experiment, sufficient benzoic acid to produce the same amount of CO_2 is burnt in the presence of GeO_2. Rotation of the calorimeter produces a final solution identical with that of the original experiments. From a thermochemical point of view the procedure eliminates the need to know the exact nature of the final solution.

Another relatively recent type of bomb calorimetry involves the combustion of a compound in a high-pressure atmosphere of a gas other than oxygen,

The enthalpies of formation of a number of fluorides have been determined in this way by combustion in fluorine gas.

2.5.4 Differential Scanning Calorimetry and Thermal Analysis

Many inorganic compounds, particularly complexes, become unstable as their temperature is raised and decompose quantitatively to well-defined products. Values of ΔH for decomposition processes can be rapidly measured by differential scanning calorimetry or by some other type of quantitative thermal analysis. In one design of the former type of equipment a small liquid or solid sample (2–30 mg) and a reference are heated at a steady rate. A closed-loop negative feedback electronic circuit controls the power to the sample and reference. An endothermic or exothermic reaction in the sample causes more or less power to be supplied to the sample to maintain equality of temperature with that of the reference, and this power is recorded as the ordinate on a chart with temperature (or time) as the abscissa. The area under the peak is thus a measure of ΔH involved in the change in the sample.

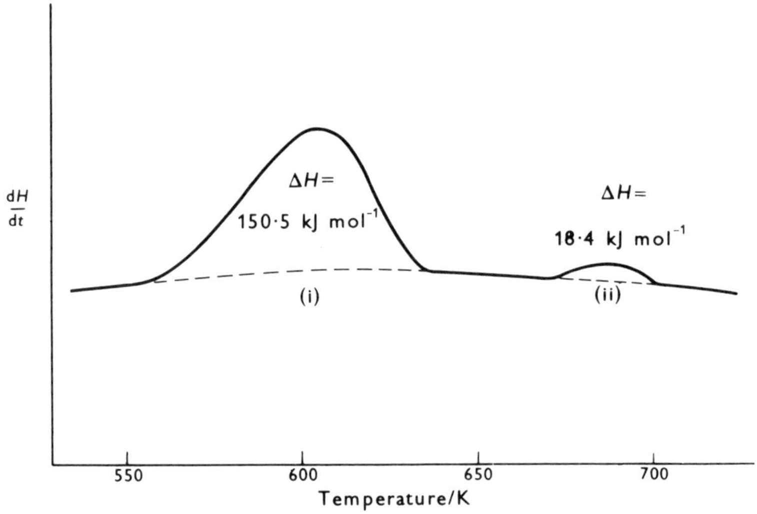

FIG. 2.4. Differential scanning calorimetric recording and values of ΔH for thermal reactions of sodium dihydrogen orthophosphate:
(i) $2Na_2H_2PO_4(s) = Na_4P_2O_7(s, III) + H_2O(g)$
(ii) a structural change; $Na_4P_2O_7(s, III) = Na_4P_2O_7(s, II)$.

Typically the values of ΔH for the stepwise removal of a gaseous ligand from a solid complex, or for thermal condensation reactions such as those shown in Fig. 2.4 for sodium phosphates, are obtained.

In addition to reaction enthalpy measurements, a differential scanning calorimeter will supply information on heat capacities, enthalpies of phase changes and, from the shapes of decomposition peaks, the kinetics of decomposition reactions.

REFERENCES

Thermodynamic Data

[1] National Bureau of Standards, *Technical Notes* 270–3, 4, 5, Washington, D.C. (1968–71).
[2] Sillén, L. G., and Martell, A. E., *Stability Constants of Metal-ion Complexes*, Chemical Society Special Publications, London (1964).
[3] Ashcroft, S. J., and Mortimer, C. T., *Thermochemistry of Transition Metal Complexes*, Academic Press, London (1970).
[4] Christensen, J. J., and Izatt, R. M., *Handbook of Metal Ligand Heats and Related Thermodynamic Quantities*, Marcel Dekker, New York (1970).

Stability

[5] Johnson, D. A., *Some Thermodynamic Aspects of Inorganic Chemistry* Cambridge University Press, Cambridge (1968).
[6] Basolo, F., and Johnson, R., *Coordination Chemistry*, Benjamin, New York (1964).

Experimental Methods

[7] Rossotti, F. J. C., and Rossotti, H., *The Determination of Stability Constants*, McGraw-Hill, New York (1961).
[8] Rossini, F. D. (ed.), *Experimental Thermochemistry*, vol. 1, Wiley, New York (1956).
[9] Skinner, H. A. (ed.), *Experimental Thermochemistry*, vol. 2, Wiley, New York (1962).
[10] Tyrrell, H. J. V., and Beezer, A. E., *Thermometric Titrimetry*, Chapman and Hall, London (1968).
[11] Adams, G. P., Carson, A. S., and Laye, P. G., *Trans. Faraday Soc.*, **65**, 113 (1969).

Chapter 3

The Non-Transition Elements

3.1 Introduction

These elements form compounds of both covalent and ionic character and it is preferable to discuss these two classes separately.

Cations are formed most readily from the elements of Groups I, II, and III. The elements of Groups V, VI, and VII tend to form anions. The reasons for this have been discussed in Chapter 1, in terms of ionization energies and electron affinities. The ion-forming tendencies of the elements are summarized in Fig. 3.1. From this, covalent compounds should be formed from those elements which are similar to each other.

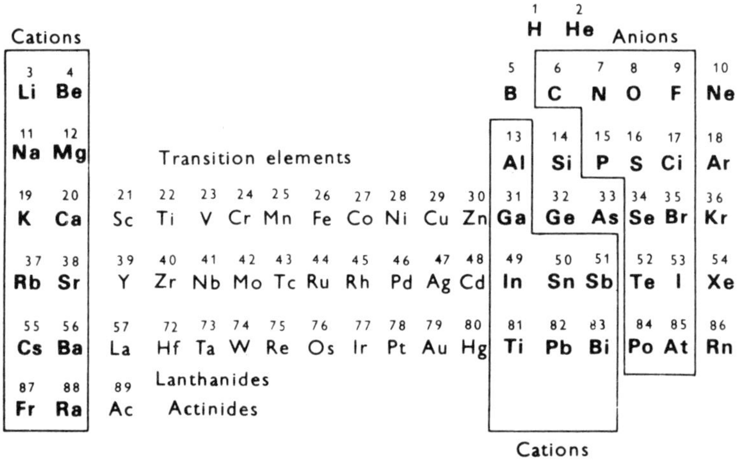

FIG. 3.1. The tendency of elements to form cations and anions. (The transition elements only form simple cations when in low oxidation states; otherwise oxycations, anions, or hydrolysed products can exist.)

Ionic compounds should be formed most easily between metals of low ionization potentials and non-metals of high electron affinity. For example, oxides and halides of Group IA and IIA metals are primarily ionic. Before discussing each class of compound, we must describe an important effect which is common to both.

3.2 The Inert Pair Effect [1, 2]

This may be defined as the tendency of the heavier elements of Groups IIIB IVB, VB (and, to a lesser extent, VIB and VIIB) to form stable valencies two less than the group valency. Therefore, a pair of electrons is expected to be neither ionized nor involved in covalent compound formation. For example, boron and aluminium normally have a valency of three whereas thallium has an extensive chemistry of the Tl^+ cation.

However, in Groups VB and beyond there is a natural restriction to the achievement of the highest (group) valency owing to valence octet restrictions. This is not related to the inert pair effect since the lone pairs in these special cases are both stereochemically and chemically active. It is still true, however, that on descending each of these groups the lone pairs become increasingly *chemically* inert pairs.

The origin of the effect lies in the energy difference between the ns^2np^x ground states and the ns^1np^{x+1} excited or valence states. The promotional energy may be estimated from atomic spectroscopy data for the process:

$$M(s^2p^x) \rightarrow M(s^1p^{x+1}) \ldots \Delta U_1$$

The *ns* and *np* orbitals must then *hybridize* to give spatially equivalent hybrid orbitals. This is illustrated in Fig. 3.2 for boron:

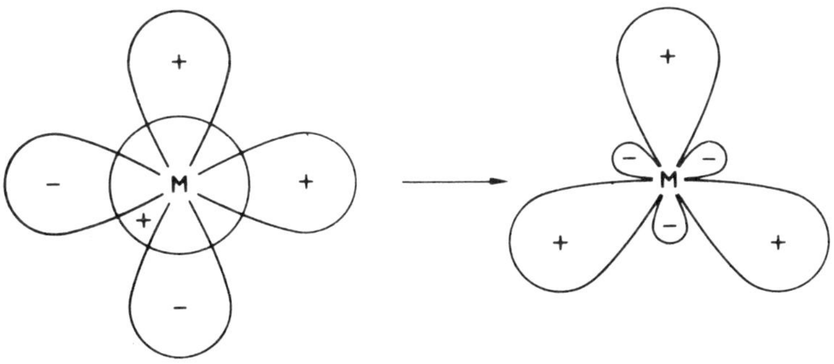

FIG. 3.2. Formation of sp^2 hybrid orbitals.

As an approximation, we shall assume that ΔU_1 contributes the major part of the excitation energy and this is summarized in Table 3.1 for the elements of Groups IIIB and IVB.

Table 3.1. Energy of the Process $M(s^2p^x) \rightarrow M(sp^{x+1})$

	B	Al	Ga	In	Tl
$\Delta U_1/kJ\ mol^{-1}$	346	349	455	418	543

	C	Si	Ge	Sn	Pb
$\Delta U_1/kJ\ mol^{-1}$	406	400	500	475	608

It can be seen that ΔU_1 *generally* increases with increasing atomic mass for these elements. The higher valencies will, therefore, only be achieved when chemical bonding forces are high enough to overcome ΔU_1. As an example, consider equations (3.1) and (3.2) relating to the formation of the Group IVB tetrahalides and dihalides in their normal physical states:

$$M(s) + 2Cl_2(g) \rightarrow MCl_4(l) \quad (3.1)$$

$$M(s) + Cl_2(g) \rightarrow MCl_2(s) \quad (3.2)$$

An enthalpy diagram may be constructed for each of the above processes, and this is shown in Fig. 3.3. We have made the assumption that $\Delta U_1 \approx \Delta H_1$.

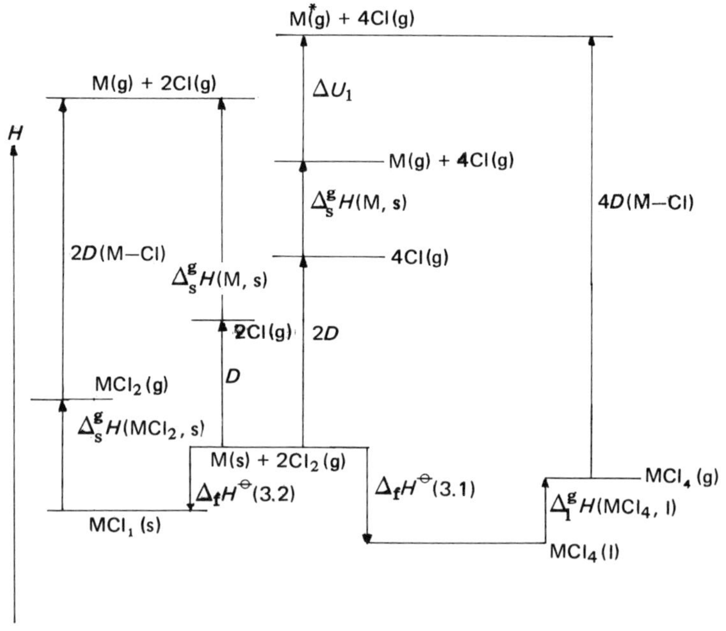

FIG. 3.3. Enthalpy diagram to illustrate the formation of di- and tetra-halides.

Covalent bond formation is represented as a common feature, even though $PbCl_2$ has a complex structure which may be only partially covalent.

From Fig. 3.3 it is clear that:

$$\Delta_f H^{\ominus}(3.1) = -[\Delta_l^g H(MCl_4, l) + 4\bar{D}(M{-}Cl)] + 2D + \Delta_s^g H(M, s) + \Delta U_1 \quad (3.3)$$

$$\Delta_f H^{\ominus}(3.2) = -[\Delta_s^g (MCl_2, s) + 2\bar{D}(M{-}Cl)] + D + \Delta_s^g H(M, s) \quad (3.4)$$

Neglecting entropy considerations, the quantity

$$\Delta H(4{-}2) = \Delta_f H(3.1) - \Delta_f H(3.2)$$

should be a rough guide to the stability of the tetrachlorides with respect to the constituent elements. $\Delta H(4{-}2)$ is given by:

$$\Delta H(4{-}2) = D + \Delta U_1 - 2D(M{-}Cl) - \Delta_l^g H(MCl_4, l) + \Delta_s^g H(MCl_2, s)$$

$D(M{-}Cl)$ will certainly differ in $MCl_4(g)$ and $MCl_2(g)$ but for numeric convenience, and in the absence of adequate data, we have assumed them to be constant. Although CCl_2 and $SiCl_2$ are very unstable, they do exist in the vapour phase; $\Delta_s^g H(MCl_2, s)$ does not, therefore, enter into the calculations for these compounds.

Some mean bond energies and enthalpies of sublimation are shown in Table 3.2. We have estimated the value of the enthalpy of sublimation of $GeCl_2$ to be 100 kJ mol^{-1} since the solid dichloride is actually ionic with the formula $Ge^+[GeCl_4^-]$.

Table 3.2. Mean Bond Dissociation Energies and Enthalpies of Sublimation of Tetrachlorides and Dichlorides of Group IVB Elements

	C	Si	Ge	Sn	Pb
$\Delta_l^g H(MCl_4, l)$/kJ mol^{-1}	33	30	34	39	38
$\Delta_s^g H(MCl_2, s)$/kJ mol^{-1}	–	–	–	146	182
$\bar{D}(M{-}Cl)$/kJ mol^{-1}	326	393	343	322	243

By using a value of $D(Cl{-}Cl)$ of 244 kJ mol^{-1} and the approximate values of ΔU_1 in Table 3.1, the values of $\Delta H(4{-}2)$ may be estimated at:

	C	Si	Ge	Sn	Pb
$\Delta H(4{-}2)$/kJ mol^{-1}	−75	−172	+24	+182	+510

The definite trend towards higher values of $\Delta H(4{-}2)$ clearly shows the decreasing thermodynamic stability of the tetrahalides.

Despite the approximations used in this model, certain features emerge:

(1) If entropy changes are neglected, as is reasonable in such comparisons, $\Delta_f H$ is a good guide to the thermodynamic stability of these compounds.

(2) With increasing atomic mass, the excitation energy of the Group IVB elements increases whilst the mean element – chlorine bond energies tend to decrease. This is very important.

(3) The sublimation enthalpies of the dihalides of higher molar mass may be particularly important with reference to the stability of Sn(II) and Pb(II).

Very similar analyses may be applied to other groups of elements and to oxides and other compounds. The same result is obtained. In all cases it is found that factor (2) is by far the most important single factor in accounting for the stability of the lower oxidation states.

To what extent is the so-called inert pair truly inert? Consider the structure of $SnCl_2$ in the gas phase:

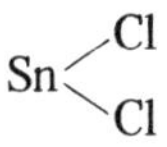

In this structure the lone (and presumed inert) electron pair is clearly active in dictating the molecular geometry. Virtually all of the structures, both in condensed and non-condensed phases, of such elements in their lower valencies show some deviation from regular geometries owing to the presence of lone pairs.

This clearly underlines an important point – an inert pair may be *chemically* inert but is usually *structurally* active. This is particularly true of Groups IIIB, IVB, and VB.

One type of compound does not fit into the regular behaviour discussed above. This is comprised of the alkyls and aryls of thallium (III) and lead (IV). Conditions which might be expected to favour the formation of the lower valent compounds invariably result in the preparation of the higher valent compounds, for example:

$$2PbCl_2\,(ether) + 4MeMgI\,(ether) \rightarrow PbMe_4\,(ether) + Pb(s) + 2MgCl_2 + 2MgI_2$$

3.3 Ionic Compounds

The theory of ionic bonding was discussed in Chapter 1 and some representative structures were discussed. The most striking aspect of ionic lattices is their high symmetry and it is this which makes possible some fairly simple, but important, calculations.

3.3.1 Lattice Energies of Ionic Compounds [3, 4]

The lattice energy (U) of an ionic compound was defined in Table 2.1 of Chapter 2 as ΔU for the conversion of a crystalline compound into the gaseous ions at infinite separation.

The forces responsible for the lattice energy are primarily electrostatic. Since any ion is closely surrounded by a number of ions of opposite charges, it is not surprising that lattice energies can amount to several thousand kilojoules per mole. It is of considerable interest to compare the results of experiment with calculated values of lattice energies. If the two are in good agreement, then calculated values may help, for example, in predicting the stabilities of hypothetical compounds.

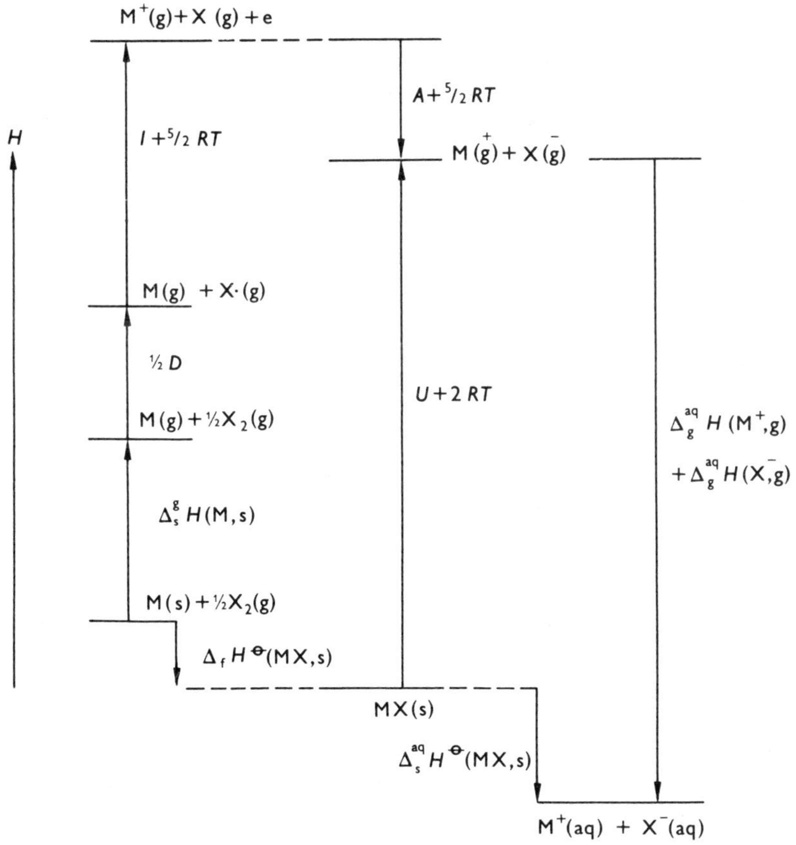

FIG. 3.4. Enthalpy diagram for the formation of MX.

3.3.2 Experimental Determination of Lattice Energies

It is not possible to measure the value of U from reaction (3.5) directly:

$$MA_n(s) \rightarrow M^{n+}(g) + nA^-(g) \qquad (3.5)$$

An indirect determination of U from other thermodynamic quantities may be achieved in two ways, as described below.

(1) *Determination of U from Sublimation Enthalpies*

Figure 3.4 shows the thermochemical cycle first proposed by Born and Haber. We illustrate its use for a binary compound MX, where M is an alkali metal and X is a halogen.

From Fig. 3.4, we find

$$U + 2RT = I + 5/2RT + 1/2D + \Delta_s^g H(M, s) - \Delta_f H^{\ominus}(MX, s) + A + 5/2RT$$

where, I = ionization potential of M(g)
D = dissociation energy of X_2(g)
$\Delta_s^g H$ = sublimation enthalpy of M(s)
A = electron affinity of X(g)

The RT terms in (3.6) are small compared to U and I, and omission of them would lead to negligible error but, in principle at least, the values of U and I refer to 0 K rather than 298 K.

This is a method which appears to be simple but has the disadvantage that electron affinities are also difficult to measure directly. Cycles of this type have, however, been widely used.

(2) *Determination of U from Hydration Enthalpies*

If we know the enthalpies of hydration $\Delta_g^{aq}H(M^+, g)$ and $\Delta_g^{aq}H(X^-, g)$ and the enthalpy of solution $\Delta_s^{aq}H(MX, s)$ of the salt, then from Fig. 3.4 we also obtain:

$$U + 2RT = \Delta_s^{aq}H(MX, s) - [\Delta_g^{aq}H(M^+, g) + \Delta_g^{aq}H(X^-, g)] \qquad (3.7)$$

Equation 3.7 is particularly useful since, even if hydration enthalpies are not known, it is often possible to estimate them from the excellent correlation between the crystal radius of the anion and $[\Delta_g^{aq}H(M^+, g) + \Delta_g^{aq}H(X^-, g)]$ as shown in Fig. 3.5 for the alkali metal halides.

Difficulties are encountered with non-spherical ions (such as the azide ion N_3^-) but, fortunately, the curvature of the corresponding graphs is slight.

Whichever method is used, it must be remembered that an *experimental* quantity results which is not dependent on any model of bonding.

Two useful applications of these thermochemical cycles are:

(1) To compare the experimental and calculated values of U. Any disparity, especially if the experimental value is the higher one, is often attributed to a covalent contribution to the bonding.

(2) To substitute a theoretical value of U into equation (3.6) in order to derive the enthalpy of formation of some hypothetical compound, as in the example in Section 2.4.

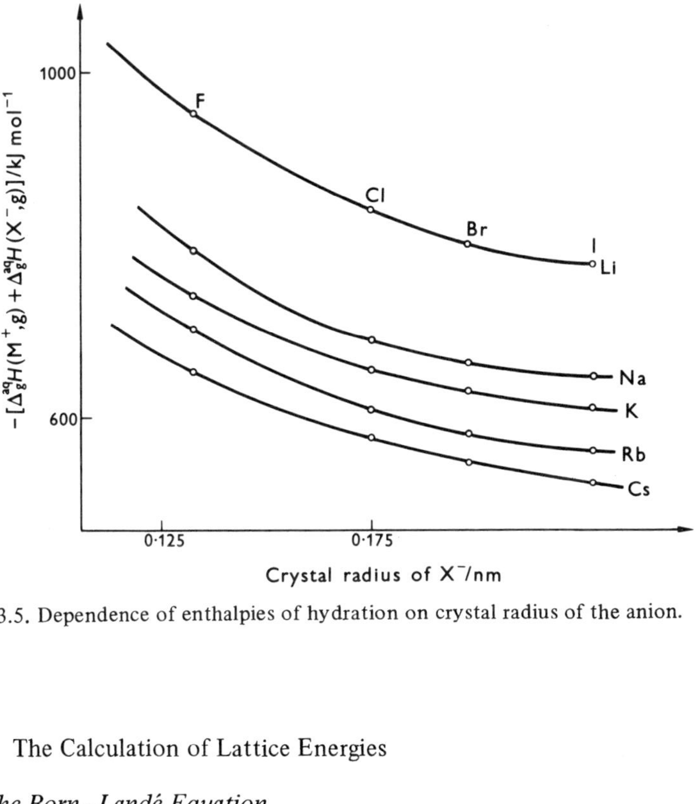

FIG. 3.5. Dependence of enthalpies of hydration on crystal radius of the anion.

3.3.3 The Calculation of Lattice Energies

(a) The Born–Landé Equation

We shall illustrate the calculation for a rock-salt lattice but the methods are of general applicability.

Each cation (charge = $z^+ . e$) is surrounded by six anions (charge = $z^- . e$) at a distance, r. This is an attractive interaction and the electrostatic energy, per cation, is given by

$$U_1 = \frac{-6z^+z^-e^2}{r \, . \, 4\pi\epsilon_0},$$

where ϵ_0 is the permittivity of free space. The next nearest neighbours are twelve cations at a distance $\sqrt{2}r$. These give rise to destabilization, and the repulsive energy is

$$U_2 = \frac{+12z^+z^+e^2}{\sqrt{2}r\,.\,4\pi\epsilon_0}$$

We then find eight anions at $\sqrt{3}r$, six cations at $2r$, and so on. This results in a series expansion for the electrostatic energy of interaction for each cation (remember that the absolute charges on the ions are equal for compounds with the rock-salt structure):

$$U_T = \frac{-z^+z^-e^2}{r\,.\,4\pi\epsilon_0}\left(6 - \frac{12}{\sqrt{2}} + \frac{8}{\sqrt{3}} - \frac{6}{2} \ldots \text{etc.}\right)$$

The series converges slowly and has been calculated with quite high precision by means of digital computers for the rock-salt and other structures. The total stabilization energy U'_T can be written, in general, as:

$$U'_T = \sum_{ij} \frac{z^+z^-e^2}{r_{ij}\,.\,4\pi\epsilon_0}$$

(where the summation is over all pairs of ions *i, j*). We denote the sum of the convergent series by the constant *M* for a given structure and write, for one mole:

$$U'_T = \frac{N_A\,.\,M\,.\,z^+z^-e^2}{r\,.\,4\pi\epsilon_0}$$

In this equation N_A is the Avogadro constant and *M* is called the Madelung constant

However, it is possible for *M* to be modified slightly such that

$$U'_T = \frac{N_A M' z^2 e^2}{r\,.\,4\pi\epsilon_0}$$

M' contains not only the geometric factor but also part of the ionic charge, z^+z^-/z^2; *z* is the highest common factor of z^+ and z^-. For example, *z* is one for NaCl, CaF_2, and Al_2O_3; *z* is two for ZnS and TiO_2. M' therefore includes the ratio z^+z^-/z^2. Textbooks almost invariably use *M* but it is more common to find complications of M' [5]. These two alternative forms of the Madelung constant are, of course, equal for all AB structures but differ for AB_2, A_2B, A_2B_3 etc. The advantage of the M' values is that they reflect the importance of both ionic charge *and* geometric structure. Values of M' are given in Table 3.3.

Table 3.3. Values of the Madelung Constant for some Common Crystal Structures

Structure	M	M'
Rock-salt	1·74756	1·74756
Caesium Chloride	1·76267	1·76267
Zinc blende	1·63805	1·63805
Wurtzite	1·64132	1·64132
Fluorite (CaF_2)	2·51939	5·03878
Rutile (TiO_2)	2·3850	4·7701
β-Quartz	2·201	4·402
Corundum (Al_2O_3)	4·040	24·242

U'_T has a stabilizing effect but we must also allow for the destabilizing effect arising from the repulsions between the electron clouds and the nuclei. This is of the form

$$U_R = +A/r^n$$

and the lattice energy must, therefore be the sum of U'_T and U_R. At an equilibrium value of $r(= r_0)$, U'_T and U_R will balance such that the lattice energy is a maximum.

Mathematically, this means that we must differentiate the total energy with respect to r in order to find the value of the constant, A. By using the notation $U_s = U'_T + U_R$, we have

$$U_s = \frac{N_A M' z^2 e^2}{r \,.\, 4\pi\epsilon_0} + \frac{A}{r^n}$$

$$\frac{dU_s}{dr} = \frac{N_A M' z^2 e^2}{r^2 \,.\, 4\pi\epsilon_0} - \frac{nA}{r^n}$$

When U_s is maximum (= U), $dU_s/dr = 0$ and $r = r_0$, the shortest cation–anion separation in the crystal. The result of this, with some manipulation, is

$$A = \frac{-N_A M' z^2 e^2}{4\pi\epsilon_0 n \,.\, r_0^{(1-n)}}$$

or,

$$U = \frac{N_A M' z^2 e^2 (1 - 1/n)}{r_0 \,.\, 4\pi\epsilon_0} \qquad (3.8)$$

n can be obtained from compressibility measurements. Fairly large errors in its value, which is usually between 5 and 10, can be shown to have only a small effect on the value of the lattice energy, U. Values of U calculated from the Born–Landé equation compare quite well with experimental results for primarily ionic compounds and some examples will be discussed later.

(*b*) *The Kapustinskii Equation*

There are certain refinements which can be made to the equation derived above, but these normally have only small effects.

It becomes increasingly more difficult to apply this apparently simple equation to more complex salts of lower symmetry and, rather than attempt to introduce more refinements, we shall now describe a different approach postulated by Kapustinskii.

Kapustinskii recognized that the Born–Landé equation could be written in the form:

$$U = \frac{N_A \nu z^+ z^- e^2 (M'/\nu)}{r_0 \,.\, 4\pi\epsilon_0}(1 - 1/n) \tag{3.9}$$

ν is equal to the number of ions in the simplest empirical formula (e.g. $\nu = 3$ for CaF_2). Although M (Table 3.3) can vary between quite wide limits, (M/ν) is much less sensitive and only varies between 0·74 and 0·88. We may regard (M/ν) as an alternative form of the Madelung constant, since its value is characteristic of any particular type of crystal structure. It is not surprising that (M/ν) is so insensitive to structure since one factor, the ionic ratio, has been eliminated. Therefore, (M/ν) is dependent purely on the coordination numbers of the anion and cation. When we consider that r_0 (the minimum cation–anion distance) increases slightly with increasing coordination number, then we finally find that there is hardly *any* variation to be expected in the ratio $(M/\nu \,.\, r_0)$ which occurs in equation (3.9). Kapustinskii postulated that the equation (3.10) could be used for any ionic compound, regardless of its ionic structure:

$$U/\text{kJ mol}^{-1} = \frac{108\, z^+ z^- \nu}{(r^+ + r^-)} \tag{3.10}$$

where r^+ and r^- are the radii (in nanometres) of the cation and anion respectively.

This surprisingly simple result has been obtained from (3.9) by using a value of nine for n and the Madelung constant of the rock-salt lattice. The use of (3.10) is, therefore, tantamount to saying that all ionic compounds are of the rock-salt type, which is clearly untrue. The reason for the simplicity is the constancy of the ratio (M/ν), a fact which holds true for a very wide range of simple and complex compounds. Further refinements of this equation are possible.

The important uses of the Kapustinskii equation are firstly, to obtain approximate values of U and, secondly, to obtain estimates of r^+ and r^- from thermochemical data. The values thus obtained are termed 'thermochemical radii'. For example, we may wish to calculate the radius of the sulphide ion in the Wurtzite form of zinc (II) sulphide. U(ZnS) is 3404 kJ mol^{-1} and $r(Zn^{2+})$ is 0·072 nm. From equation (3.10) we find

$$r(S^{2-}) = \{[(108{\cdot}0 \,.\, 2^2 \,.\, 2)/3404] - 0{\cdot}072\} \text{ nm}$$
$$= 0{\cdot}181 \text{ nm}$$

This compares rather well with the literature value of 0·184 nm obtained from X-ray measurements.

If the lattice energies are not known (this is the case for many coordination compounds) then it is still possible to obtain the thermochemical radius of one of the ions. Suppose that we have two complex salts with the same complex anion (A) but different simple cations, M_1 and M_2. Then, by considering simple thermochemical cycles it is easily shown that

$$U(M_1A) - U(M_2A) = \Delta_f H^{\ominus}(M_1^{z+}, g) - \Delta_f H^{\ominus}(M_2^{z+}, g) - \Delta_f H^{\ominus}(M_1X, s) - \Delta_f H^{\ominus}(M_2X, s)$$

and, numerically, from (3.10):

$$U(M_1A) - U(M_2A) = 108{\cdot}0 \,.\, z^+ \,.\, z^- \,.\, \nu\left[\frac{1}{r(M_1^{z+}) + r(A^{z-})} + \frac{1}{r(M_2^{z+}) + r(A^{z-})}\right] \text{ kJ mol}^{-1}$$

From this equation we can calculate $r(A^{z-})$, provided that the other parameters are known.

One of the few objections we could make to the Kapustinskii approach is that, in some cases, the lattice spacing is not always governed by cation–anion contacts but sometimes by anion–anion contacts. This is the case, for example, in the sodium iodide lattice.

There are more refined methods of approaching the problem but the simple one outlined above is sufficientlv accurate to give useful results.

3.4 Applications of Lattice Energy Measurements

3.4.1 Thermochemical Properties of the Oxides, Peroxides, and Superoxides of Group IA and IIA Metals

These compounds present us with an interesting series, in that the reactions of many metals with oxygen invariably result in the formation of one or more members of the series. Oxides are widely distributed but peroxides and superoxides are more well known for the IA and IIA metals; even in these groups, there is considerable variation in behaviour. For example, lithium reacts with oxygen to form the monoxide (Li_2O) whereas sodium and potassium form the peroxide (Na_2O_2) and the superoxide (KO_2) under similar conditions. It is of interest to attempt to understand the reasons for this behaviour in terms of the thermochemical properties of the compounds.

Some relevant data for the ions are:

	O^{2-}	O_2^-	O_2^{2-}
Ionic radii/nm	0·135	0·163	0·180
$\Delta_f H^\ominus$/kJ mol^{-1}	+925	−79	+715

(Note that, for O_2^- and O_2^{2-}, thermochemical radii are quoted.) Some enthalpies of formation and lattice energies are summarized in Table 3.4.

Table 3.4. Thermochemical Properties of Group IA and IIA Oxides and Related Compounds

Compound	$\Delta_f H^\ominus$/kJ mol^{-1}	U/kJ mol^{-1}
Li_2O	−594	2908
Na_2O	−414	2519
K_2O	−276	2230
Rb_2O	−240	2146
BeO	−634	4527
MgO	−674	3925
CaO	−690	3519
SrO	−613	3314
BaO	−561	3121
Li_2O_2	−636	2715
Na_2O_2	−506	2425
K_2O_2	−494	2222
Rb_2O_2	−427	2113
Cs_2O_2	−401	2019
MgO_2	−623	3664
CaO_2	−661	3075
SrO_2	−644	2920
BaO_2	−632	2707
NaO_2	−259	782
KO_2	−284	720
RbO_2	−289	703
CsO_2	−297	686

The formation of the compounds is analysed in terms of the thermochemical cycles in Fig. 3.6. for the reactions of the alkali metals. Similar cycles can be constructed for the alkaline Fig. 3.6 earth salts.

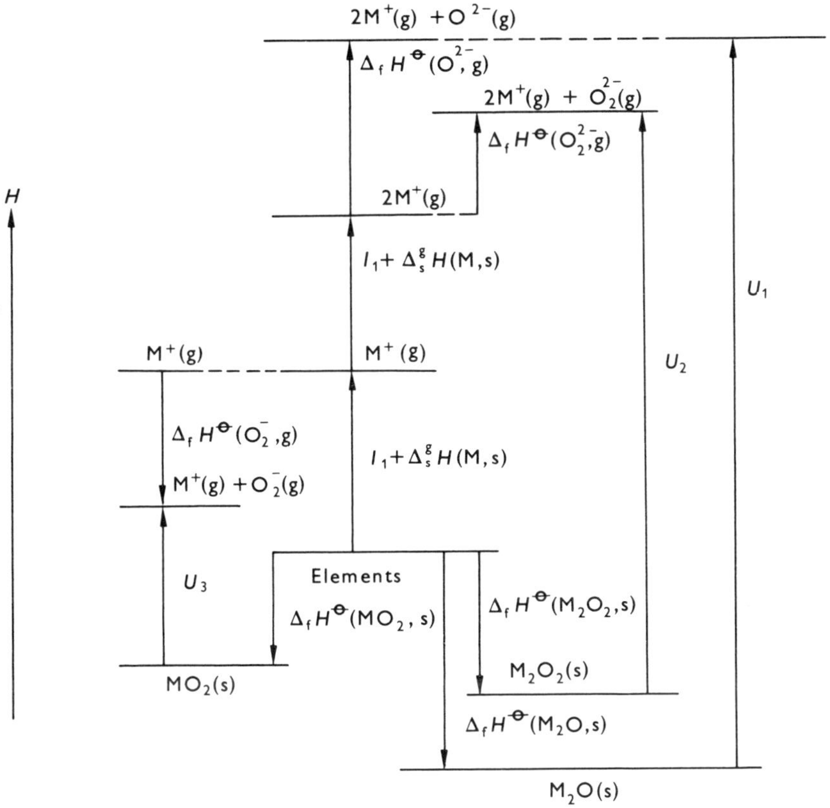

FIG. 3.6. Enthalpy diagram for the formation of the oxides, peroxides, and superoxides of the alkali metals.

If we assume that $U \approx \Delta H(298{\cdot}15\ \mathrm{K})$, we obtain

$$\Delta_f H^\ominus(M_2O, s) = -U_1 + 2I_1 + 2\Delta_s^g H(M, s) + \Delta_f H^\ominus(O^{2-}, g) \quad (3.11)$$

$$\Delta_f H^\ominus(M_2O_2, s) = -U_2 + 2I_1 + 2\Delta_s^g H(M, s) + \Delta_f H^\ominus(O_2^{2-}, g) \quad (3.12)$$

$$\Delta_f H^\ominus(MO_2, s) = -U_3 + I_1 + \Delta_s^g H(M, s) + \Delta_f H^\ominus(O_2^-, g) \quad (3.13)$$

Similar results are again obtained for the alkaline earth salts. With reference to Table 3.4 and equations (3.11–3.13), we can detect several interesting trends:

(i) For a given type of compound, U decreases on descending a group; this is due to the increasing size of the cation which, in turn, increases the denominator in (3.8).

(ii) In any period, U increases in the order

$$M^{I}O_2 \ll M^{I}_2O_2 < M^{I}_2O < M^{II}O_2 < M^{II}O$$

This sequence reflects both size and charge factors. The high values for $M^{II}O$ (cryst) are due to the small size of the cation and the favourable value of $(z^+ . z^-)$ in (3.8).

(iii) $\Delta_f H^{\ominus}(O_2^-, g)$ is negative so that this should favour the formation of superoxides in all cases. However, the small charge on the ion and its large radius give rise to small lattice energies (Table 3.4). Therefore, only those elements with small values of $[I_1 + \Delta_s^g H(M^+, g)]$ will form superoxides. These elements include the heavier elements of Group IA.

(iv) Enthalpy factors, though important, are not always dominant. For example, lithium monoxide is formed in preference to the peroxide owing to the smaller loss of entropy. Similarly, the superoxides form the peroxides as the first decomposition product on heating:

$$2MO_2(s) \rightarrow M_2O_2(s) + O_2(g)$$

On descending Group IA, this reaction becomes increasingly less favourable according to the enthalpy changes calculated from Table 3.4. It proceeds because of the positive $T\Delta S$ term. We may also note the very high values of $U(Li_2O$, cryst) and U(BeO, cryst) which are attributed to the anomalously small cations.

3.4.2 Lattice Energies of Univalent and Divalent Metal Halides

The calculated values of the lattice energies of the alkali metal halides are in remarkably good agreement with theoretical calculations, even at a low level of sophistication. Table 3.5 illustrates this point for the chlorides of the alkali metals.

Table 3.5. Lattice Energies of Group IA Chlorides (U/kJ mol^{-1})

Compound	Experimental	Born Eq.	Kapustinskii
LiCl	829	803	803
NaCl	770	757	753
KCl	699	690	682
RbCl	678	661	661
CsCl	640	623	623

Clearly, the simple theoretical models work extremely well in these cases.

Such good agreement is not always observed. This is shown by the results in Table 3.6 for silver, thallium, and copper halides. The agreement is very poor in some cases, and the difference between experimental and calculated values increases in the order;

$$F < Cl < Br < I$$

This is the order of increasing polarizability and decreasing electronegativity of the halogens and is, therefore in the same order as the expected importance of covalency in the salts. The covalent bonding must, therefore, supplement the ionic bonding in the order indicated. Also, the deviations for a given halide are always in the order:

$$Tl < Ag < Cu$$

and this is the order of increasing ionization potentials and decreasing ionic radius. It is, therefore, also the order in which covalency effects would be expected to increase in importance.

In some cases, it has been shown that our simple models of bonding are not adequate and a more complete theory would have to take account of variable degrees of covalency, amongst other factors. In particular, the transition metal compounds give very interesting results, but we defer discussion of them until a later chapter.

Table 3.6. Lattice Energies of Univalent Metal Halides (U/kJ mol^{-1})

Compound	Calc. (Kapustinskii)	Experimental
AgF	870	966
AgCl	782	916
AgBr	757	908
AgI	736	896
TlCl	665	736
TlBr	648	724
TlI	619	703
CuCl	862	979
CuBr	828	971
CuI	787	958

3.4.3 The Stability of Oxidation States [6]

The higher oxidation states of metals are usually stabilized in the form of oxides and fluorides. The reasons that fluorine stabilizes higher oxidation states are:

(a) The F—F bond dissociation energy is small (158 kJ mol^{-1}) compared with the other halogens (D(Cl—Cl) = 243 kJ mol^{-1})

(b) the first factor, combined with the electron affinity of fluorine yields an anomalously large negative enthalpy of formation for F^-(g).

(c) The small ionic radius of F^- gives rise to a large lattice energy.

These favourable enthalpy factors are often sufficient to overcome the high ionization potentials required to produce high valent compounds.

The halides thermally decompose according to the general equation

$$MX_n(s) \rightarrow MX_{(n-2)}(s) + X_2(g)$$

The ease of reduction of iodides then reflects the ease of formation of I_2(g) from I^-. The enthalpy changes for the reaction

$$2X^-(g) \rightarrow X_2(g) + 2e \qquad (3.14)$$

are given below:

	F	Cl	Br	I
ΔH(3·14)/kJ mol^{-1}	542	492	499	456

The reaction (3.14) is energetically more difficult for fluorine and this combined with the lattice energies of the halides results in a stabilization of the higher oxidation states in the order

$$F > Cl > Br > I$$

A similar situation exists for the oxides compared to the other chalcogenides. The enthalpy of formation of O^{2-}(g) is highly endothermic owing to the large value of D(O—O) and to the very unfavourable reaction,

$$e + O^-(g) \rightarrow O^{2-}(g)$$

As a result of these factors, we find that $\Delta_f H^\ominus(O^{2-}, g)$ is + 925 kJ mol^{-1} from which we would expect that oxides would be less stable than the analogous fluorides of the same metal in the same oxidation state. The reason for the stability of high valence oxides can not, therefore, be connected with the stability of the anion which was one reason for the formation of the fluorides. We focus attention instead on the favourable lattice energy predicted by the appropriate forms of the Kapustinskii equations,

$$U(MO_{n/2})/\text{kJ mol}^{-1} = \frac{108{\cdot}0 \,.\, n \,.\, (n+2)}{r(M^{n+}) + r(O^{2-})}$$

$$U(MF_n)/\text{kJ mol}^{-1} = \frac{108{\cdot}0 \,.\, n \,.\, (n+1)}{r(M^{n+}) + r(F^-)}$$

The larger radius of the oxide ion generally gives rise to smaller lattice energies but this is offset by small values of n. The result is that high valencies are more thermally stable in fluorides. For example, we may compare HgO and HgF_2 : HgO decomposes thus

$$HgO(s) \rightarrow Hg(l) + 1/2\ O_2(g)$$

but the fluoride merely vaporizes unchanged.

3.5 Covalent Compounds

3.5.1 Bond Energies

The distinction between bond dissociation energies, coordinate bond dissociation energies, and bond energy terms has been made in Chapter 2. We shall now discuss the thermochemical properties of covalent compounds in terms of bond dissociation energies. The thermochemistry of covalent polymeric compounds is treated in Chapter 6.

3.5.2 Energies of Single Bonds

It is not always possible to say that a molecule contains a set number of single and multiple bonds, since experimental methods cannot always be relied upon to provide a distinction. For example, the B–F bond length in BF_3 is appreciably shorter than that in BF_4^-. However, the extent to which this is due to π-bonding or to the expected closer packing in a trigonal arrangement cannot be ascertained.

To be correct, we should be interested not in the mean thermochemical bond dissociation energies but, rather, in intrinsic bond dissociation energies. These refer to the dissociation of the compound into its atoms, in the gas phase, such that all of the atoms are in their valence, rather than ground states. In practice, however, we are normally restricted to ground states, and bond dissociation energies will be discussed in this more limited sense. A further discussion of this subject is given in Section 5.1.

A useful estimate of heteronuclear mean bond dissociation energies is given by

$$\bar{D}(\text{A–B}) = \tfrac{1}{2}[D(\text{A–A}) + D(\text{B–B})] + 96{\cdot}8(x_A - x_B)^2 \qquad (3.15)$$

All of the bond energies are in kJ mol^{-1} ; x_A and x_B are the Pauling electronegativity coefficients, which measure the electron attracting power of an atom in a molecule.

As an example of the application of (3.15), let us compare the calculated values of $\bar{D}$(M–Cl) with the experimental values for the gaseous tetrahalides (NBS Technical Note 270–3); D(M–M) values refer to gaseous M_2R_6 and are taken from Chapter 5 or are estimated by differences. In the table, below, x_M is given in parentheses after each element:

		C(2·55)	Si(1·90)	Ge(2·01)	Sn(1·96)	Pb(2·33)
$\frac{1}{2}[D(\text{M–M}) + D(\text{Cl–Cl})]$/kJ mol		257	253	250	231	213
$\bar{D}$(M–Cl) /kJ mol^{-1}	Experimental	327	400	340	315	243
	eqn. (3.15)	293	407	378	370	280

In this case, agreement is seen to be reasonable. This is not always true – for example, according to (3.15), the difference between $\bar{D}$(M–X) and $1/2\,[D(\text{M–M}) + D(\text{X–X})]$ should be about the same for the third and successive members of a group if Pauling electronegativity values are used. This is not generally true, and somewhat better results are obtained if the Allred–Rochow scale of electronegativity coefficients are used. It should be remembered, however, that bond energies are sensitive functions of overlap integrals of the form

$$S_{AB} = \int \psi_A \psi_B \, dv$$

and it is to be expected that generalizations may not be particularly successful in certain cases.

3.5.3 Multiple Bonding

The formation of multiple bonds, particularly π-bonds, can be established with certainty in a number of cases. Multiple bonds are formed as a result of 'sideways' overlap, as discussed in Chapter 1, by using appropriate atomic orbitals. However, since orbital overlap is clearly more effective *along* the bond axis, then π-bonding is expected to be less effective than σ-bonding in any circumstance.

Criteria used to establish the relative importance of π-bonding have included:

(1) *Measurements of bond lengths and bond angles:* the formation of a π-bond shortens bond lengths between first period elements by about 0·01 nm and increases bond angles by several degrees in those cases where planarity improves the effectiveness of π-bonding.

(2) *Spectroscopic measurements:* evidence from this source is diverse in nature but often speculative. An often cited example is the shift to lower frequencies of the C–O band in the infrared spectra of metal carbonyls. This is attributed to the back-donation of electronic charge from the metal atom to the anti-bonding orbitals of the carbon monoxide ligand molecules (p. 113).

(3) *Thermochemical measurements:* measurement of the enthalpies of formation of a series of related compounds can often be used to indicate the progressive importance of π-bonding, if present.

It should be remembered that the methods are only really conclusive when π-bonding is very strong; in other cases more speculative arguments result. Furthermore, energetically weak π-bonding may be of considerable structural importance. An example of this is found again in the metal carbonyls where the main function of π-bonding is to remove excess negative charge from a central metal atom and the consequent slight strengthening of the σ-bonding is only incidental.

3.5.4 Types of π-Bonding

(a) $p_\pi - p_\pi$

Figure 3.7 illustrates the radial dependence of the p-orbitals on n the

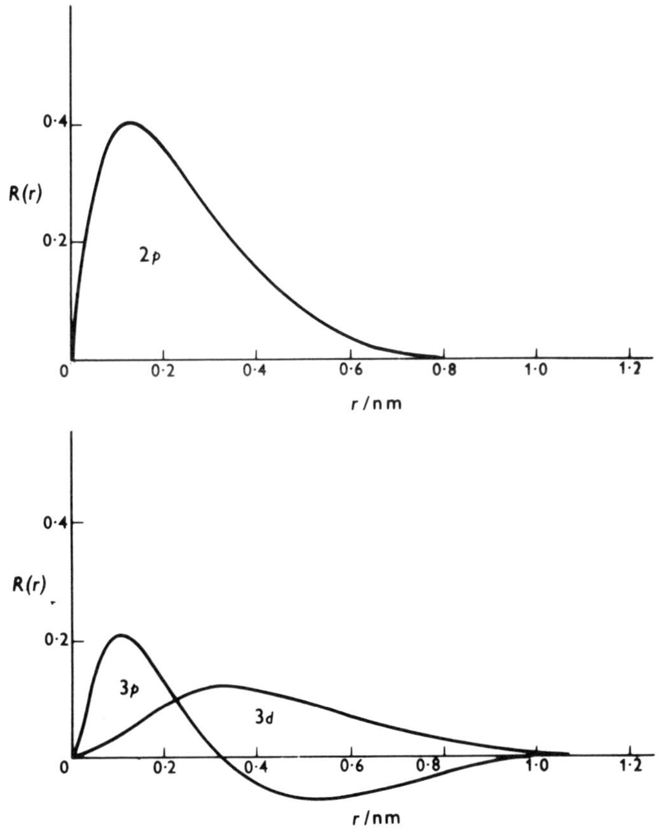

FIG. 3.7. The radial parts, $R(r)$, of the $2p$, $3p$, and $3d$ orbitals as a function of the distance from the nucleus, r.

principal quantum number. An increase in n, from two to three, results in an increase in diffuseness of the orbitals. This indicates that $p_\pi - p_\pi$ bonding is of greatest importance for bonds between first period elements and is of negligible importance for homonuclear bonds between elements of the third and subsequent periods. Thus, the overlap integral

$$S = \int \pi_p \pi_p \, dv$$

is 0·217 in O_2 and only 0·177 in S_2. Furthermore, although CO_2 and CS_2 exist, CSe_2 and CTe_2 are not stable species.

(b) d_π-bonding

Figure 3.7 also illustrates the radial dependence of the 3*d* orbital. The greater diffuseness of the *d*-orbitals is responsible for the weakness of d_π-bonding. In tetrahedral oxyanions and certain cyclic systems $d_\pi - p_\pi$ bonding is important structurally and this will be dealt with in a later chapter (p. 139).

We shall now discuss the thermochemical properties of some classes of covalent compounds.

3.6 Thermochemical Properties of Covalent Compounds

3.6.1 Addition Compounds of Group IIIB Halides and Alkyls [7]

These compounds react with inorganic and organic bases to form stable adducts;

$$EX_3 \text{ (planar)} + L: \rightarrow L: EX_3 \text{ (tetrahedral)}$$

The base L is an electron donor such as an amine or ether. Some fairly typical enthalpies of addition in solutions using non-coordinating solvents are given in Table 3.7. The enthalpies of addition refer to the reactants and products in their ground states. In order to compare the intrinsic bond dissociation energies, we should correct the measured enthalpies to allow for the excitation of the atoms to their valence states and for the changes in orbital coefficients for the process:

$$EX_3 \text{ (planar)} \rightarrow EX_3 \text{ (trigonal)}$$

The nature of these corrections is only approximate owing to our imprecise knowledge of the magnitudes of the atomic changes involved. Within a series (such as that in Table 3.7) it would seem better to consider variations in enthalpy changes rather than absolute values.

The enthalpies of addition of aliphatic amines, $Me_nH_{3-n}N$, to BMe_3 are in the order

$$NH_3 < MeNH_2 < Me_2NH > Me_3N$$

This reflects (i) the increasing basicity of the amines, due to the electron-releasing methyl groups and (ii) the steric crowding due to the three methyl groups in trimethylamine. The variation in bond energies is primarily due, therefore, to changes in σ-bond energies. A similar explanation applies to the pyridine and 2-methylpyridine adducts. Although the latter ligand is more basic, the 2-methyl group weakens the B–N bond by a steric interaction.

Table 3.7. Enthalpies of Addition $\Delta_a H$ of Bases to Group IIIB Halides and Alkyls

Base	EX_3	$-\Delta_a H$/kJ mol^{-1}
NH_3	BMe_3	57·5
$MeNH_2$	BMe_3	73·8
Me_2NH	BMe_3	80·6
pyridine	BMe_3	64·0
2-methyl pyridine	BMe_3	41·6
pyridine	BF_3	143
pyridine	BCl_3	189
pyridine	BBr_3	217
pyridine	$AlCl_3$	288
pyridine	$AlBr_3$	265
pyridine	AlI_3	265
Me_3N	BF_3	129
Me_3P	BF_3	79
Me_3N	BMe_3	73·6
Me_3P	BMe_3	68·9
Me_3N	$GaMe_3$	88
Me_3P	$GaMe_3$	75

The series BX_3/pyridine is interesting in that we might expect, from the relative electronegativities of the halogens, that the greatest induced positive charge should be on the boron in BF_3. Thus the reaction,

$$BF_3 + L: \rightarrow L\ BF_3 \ldots \Delta_a H$$

might be expected to have the highest $\Delta_a H$. The inversion (Table 3.7) may be explained by postulating partial $p_\pi - p_\pi$ bonding in the B–X bonds:

This is, essentially, a way of removing excess of negative charge from the halogen to the empty boron p_z orbital. In view of the electronegativities of the halogens, this should be in the order

$$F > Cl > Br$$

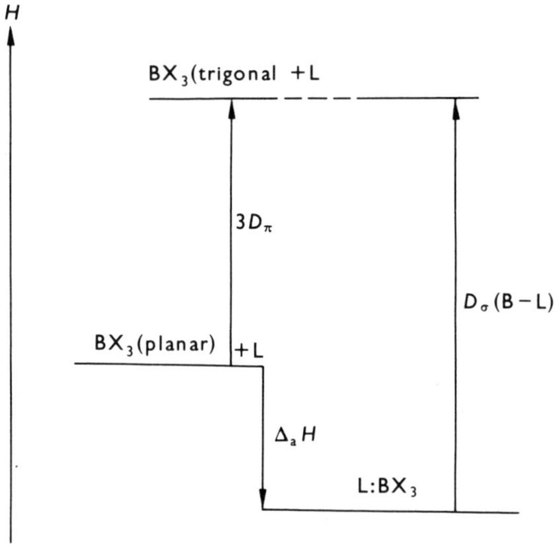

FIG. 3.8. Enthalpy diagram for the formation of boron halides adducts.

Adduct formation could be represented as a series of energetic steps, as shown in Fig. 3.8. If the increase in D_σ(B–L) for the series X = F, Cl, Br is *smaller* than the decrease in $3D_\pi$, then the inversion of the series is explained. The absence of such a marked effect for the aluminium halides may reflect the decreased importance of $p_\pi - p_\pi$ bonding.

Of course, it is quite possible to derive an alternative argument based on halogen–halogen repulsions. On forming the BF_3 adduct, the destabilization resulting from bringing the fluorine atoms closer together during adduct formation could be sufficient to cause the observed reversed trend but no precise calculations of this effect have been reported.

In the sterically-unhindered systems L: BF_3, we find $\Delta_a H$ to be

$$Me_3N > Me_3P$$

For the analogous BMe_3 adducts the difference is much smaller, reflecting the greater steric crowding in the trimethyl system. The difference in $\Delta_a H$ again increases when boron is replaced by gallium where steric crowding is reduced by the larger central atom.

3.6.2 Covalent Hydrides

The covalent hydrides display a wide range of structural behaviour and it is of interest to see if there are energetic explanations for their properties. The types of structure commonly encountered are:

(i) Simple monomeric (possibly associated) molecules.

(ii) Catenated molecules – those in which there are element-to-element bonds in addition to element-hydrogen bonds. An example is the C_nH_{2n+2} series of alkanes.

(iii) Bridged structures containing multicentre E–H–E bonds, in addition to terminal E–H bonds. The best-known example of such bonding occurs in the boron hydrides, the simplest representative being diborane B_2H_6. This molecule contains four B–H terminal bonds and two three-centre/two-electron B–H–B bridging bonds:

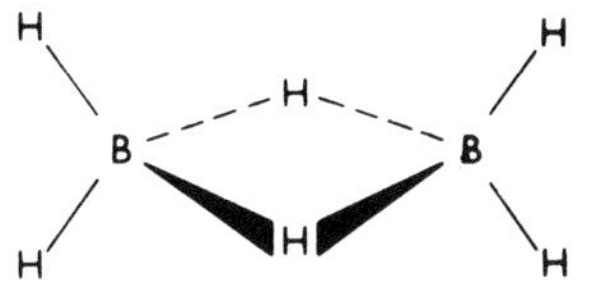

The variation in structure is fairly systematic, as can be seen in Fig. 3.9. We shall now discuss the energetics of some representative classes of hydrides.

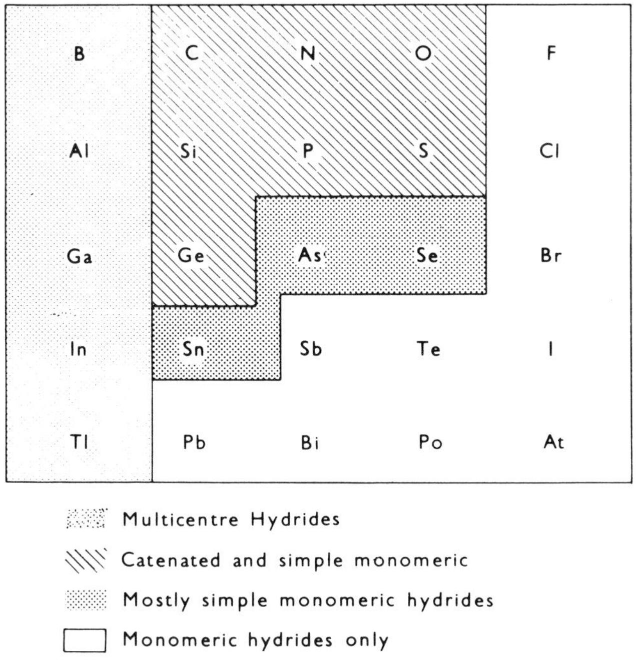

Fig. 3.9. Occurrence of the various types of hydrides of Groups IIIB to VIIB.

3.6.3 Group IIIB Hydrides [8]

The boron hydrides mentioned above are 'electron deficient' because of their unusual multicentre bonds. The simplest boron hydride might be expected to be much simpler than these electron deficient compounds and to have the formula BH_3. This molecule (borine) is difficult to isolate even though derivatives of the formula L: BH_3 are known. The reason for this is that there is no mechanism in the BH_3 molecule whereby electron density can be delocalized from the hydrogen atoms to the boron. The electronegativity of boron (B < H) results in a nett excess of positive charge being placed on the boron which can be removed by adduct formation (see Section 3.5.1) or by dimerization:

$$2BH_3 \rightarrow B_2H_6$$

An upper spectroscopic limit for D(B–H) is 339 kJ mol^{-1}. $\Delta_f H^\ominus(BH_3, g)$ can then be estimated as,

$$\Delta_f H^\ominus(BH_3, g) = \Delta_f H^\ominus(B, g) + 3/2D(H\text{–}H) - 3D(B\text{–}H)$$

from which we calculate

$$\Delta_f H^\ominus(BH_3, g) = +200 \text{ kJ mol}^{-1}$$

Since $\Delta_f H^\ominus(B_2H_6, g) = 35{\cdot}6$ kJ mol^{-1}, we can estimate that for the reaction

$$B_2H_6(g) \rightarrow 2BH_3(g)$$

$\Delta H^\ominus$ is approximately 364 kJ mol^{-1}. Thus the non-classical bonds, although unusual, are of considerable stability with respect to the elements. D(B–H–B) is roughly half D(B–H) which is the expected magnitude since there are only two bonding electrons to be shared between three nuclei.

The other hydrides of this group are polymeric. Little useful thermochemical data are available.

3.6.4 Group IVB Hydrides

The unique ability of carbon to undergo catenation is responsible for the great stability of compounds with C_n backbones, as in the paraffins C_nH_{2n+2}. Silicon has less tendency to catenate, the highest value of n being about 8. Higher values have been found in the catenated halides of silicon but some of the structures are in doubt. Germanes up to Ge_9H_{20} exist but the highest stannane is Sn_2H_6 and the only plumbane is PbH_4. Pronounced catenation is therefore limited to the first three members of this group.

The values of D(E–E) are as follows:

E =	C	Si	Ge	Sn
D(E–E)/kJ mol^{-1}	348	194	157	143

D(Sn–Sn) is estimated from the enthalpy of atomization of metallic tin. There is some doubt regarding D(Si–Si) since electron impact values are considerably greater than normal calorimetrically-derived values. However, recent enthalpies of explosion [9] yield data of good internal consistency, as shown in Table 3.8.

Table 3.8. Bond Dissociation Energies of Si–Si and Si–H Bonds

Compound	$\Delta_f H$(g)/kJ mol^{-1}	D(Si–H)/kJ mol^{-1}	D(Si–Si)/kJ mol^{-1}
SiH_4	30·5	320	–
Si_2H_6	71·5	320	194
Si_3H_8	108	320	196

Silanes invariably react by Si–H cleavage but this indicates that kinetic, rather than thermodynamic factors, control such reactions.

When discussing stability, it is important to specify the reaction precisely.

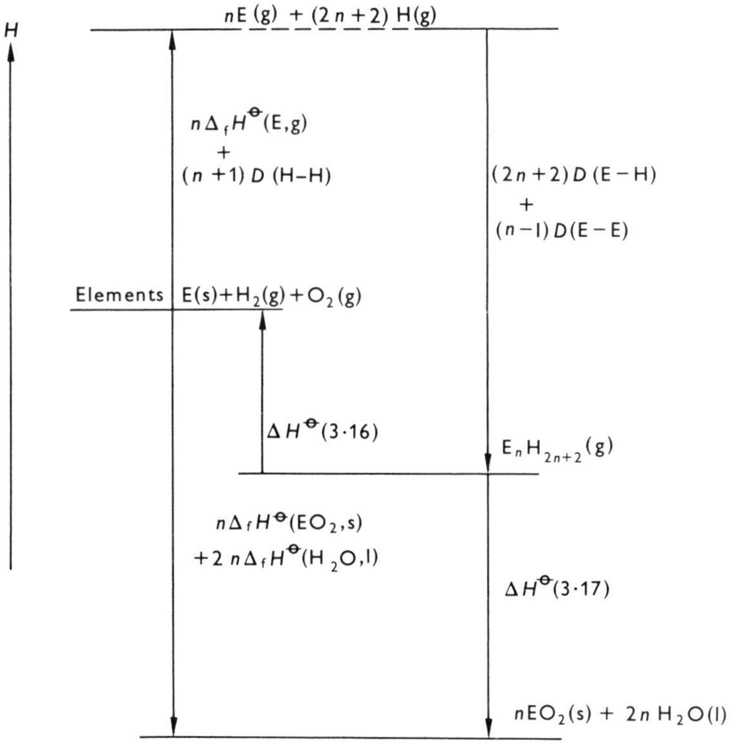

FIG. 3.10. Enthalpy diagram for reactions (3.16) and (3.17).

For example, thermal decomposition (3.16) or oxidative degradation (3.17) are possible criteria:

$$E_nH_{2n+2}(g) \rightarrow nE(s) + (n+1)H_2(g) \quad (3.16)$$

$$E_nH_{2n+2}(g) + (2n+1)O_2(g) \rightarrow nEO_2(s) + (n+1)H_2O(g) \quad (3.17)$$

In order to calculate the enthalpies of these reactions, $\Delta H^{\ominus}(3.16)$ and $\Delta H^{\ominus}(3.17)$, we require certain ancillary data (Fig. 3.10)

$$\Delta H^{\ominus}(3.16) = (2n+2).D(\text{E–H}) + (n-1)D(\text{E–E}) - n\Delta_f H^{\ominus}(\text{E, g}) - (n+1)D(\text{H–H})$$

$$\Delta H^{\ominus}(3.17) = n\Delta_f H^{\ominus}(\text{MO}_2\text{, s}) + 2n\Delta_f H^{\ominus}(\text{H}_2\text{O, 1}) - n\Delta_f H^{\ominus}(\text{E, g}) - 2nD(\text{E–H}) + (n-1)D(\text{E–E})$$

The results are summarized in Table 3.9. Although $\Delta H(3.17)$ predicts instability for silanes, it is much less successful for germanes and stannanes. Clearly, kinetic factors are more important in determining oxidative stability but $\Delta H(3.16)$ is a better guide since the reagent has been specifically excluded. Experimentally, the order of oxidative stability is

$$\text{C} \gg \text{Si} < \text{Ge} > \text{Sn} > \text{Pb}$$

Table 3.9. Values of (a) $+\Delta H^{\ominus}(3.16)/\text{kJ mol}^{-1}$ **and** (b) $-\Delta H^{\ominus}(3.17)/\text{kJ mol}^{-1}$

$n =$	1		2		3		4		5	
	(a)	(b)	(a)	(b)	(a)	(b)	(a)	(b)	(a)	(b)
C	459	1366	570	2384	1029	3402	1488	4420	1947	5438
Si	378	1435	562	2676	940	3917	1318	5158	1696	6399
Ge	359	933	561	1709	920	2485	1279	3261	1638	4037
Sn	344	812	575	1481	934	2150	1293	2819	1652	3488

3.6.5 Group VB Hydrides

There is considerably less tendency to catenation in this group. Hydrazine is the only such compound of considerable stability; P_2H_4 is rather stable but tetrazan, $NH_2(NH)_2NH_2$ is of very low stability. On descending the group, even the hydrides EH_3 become thermally unstable. The reason for this is the steady decrease in $D(\text{E–H})$ and $D(\text{E–E})$:

	N	P	As	Sb	Bi
$D(\text{E–H})/\text{kJ mol}^{-1}$	389	326	297	~260	~230
$D(\text{E–E})/\text{kJ mol}^{-1}$	159	209	180	142	~100

$D(\text{N–N})$ is anomalously low owing to lone pair repulsions.

The enthalpies of formation of the hydrides are given by

$$\Delta_f H^\circ(EH_3, g) = \Delta_f H^\circ(E, g) + 3\Delta_f H^\circ(H, g) - 3D(E\text{–}H)$$

$$\Delta_f H^\ominus(E_nH_{n+2}, g) = n\,\Delta_f H^\ominus(E, g) + (n+2)\,\Delta_f H^\ominus(H, g) - (n-1)D(E\text{–}E) - (n+2)D(E\text{–}H)$$

The experimental and calculated values for EH_3 and E_2H_4 are

	N	P	As	Sb	Bi	
$\Delta_f H^\ominus(EH_3, g)$/kJ mol^{-1}	−46·1	5·4	66·4	145	–	(exptl.)
	−40·4	−9·0	65·4	136	171	(calcd.)
$\Delta_f H^\ominus(E_2H_4, g)$/kJ mol^{-1}	95·4	20·9	–	–	–	(exptl.)
	103	−12	109	215	266	(calcd.)

$\Delta_f H^\ominus(E_nH_{n+2}, g)$/kJ mol^{-1} becomes even less favourable with increasing values of n. The instability of N_2H_4 with respect to N_2 reflects the strength of the bonding in the latter. The decreasingly favourable value of $\Delta_f H^\ominus(EH_3, g)$ results in the facile thermal decomposition to the elements of bismuthine.

3.6.6 Group VIB Hydrides

The strengths of the E–E and E–H bonds show a similar variation to that found in the preceding group:

	O	S	Se	Te	Po
D(E–E)/kJ mol^{-1}	142	263	159	~100	–
D(E–H)/kJ mol^{-1}	467	347	276	238	~200

Consequently, catenation is most marked for sulphur and this is a very important aspect of this element's chemistry. A common catenated form of sulphur is the rhombic allotrope which contains S_8 rings. The rings open readily on heating or in chemical reactions

e.g. $$Na_2S(aq) + n/8\, S_8(s) \xrightarrow[H_2O]{} Na_2S_{(n+1)}(aq)$$

In this example, the chain is kept apart (if n is large) by electrostatic repulsions from the chain ends.

Covalent compounds such as the hydrides easily lose sulphur on heating since the total number of S–S bonds remains constant,

e.g. $$H_2S_n(s) \rightarrow H_2S_{(n-m)}(s) + m/8\, S_8(s)$$

This reaction has a positive value of ΔS and is favoured by higher temperatures.

Catenation is limited to H_2O_2 for oxygen. Although polyselenides and polytellurides exist, the stability of the hydrides is much lower than that of the analogous sulphur compounds, as expected from the bond energy data.

3.6.7 Group VIIB Hydrides

These are, more correctly, halides and there is little tendency to catenation owing to valence-shell restrictions. Certainly, there are no catenated hydrides, or other covalent compounds.

The hydrogen halides are all acids. The strengths of the acids in water are in the order:

$$\text{HI, HBr, HCl} > \text{HF}$$

The similarity in strengths of HI, HBr, and HCl is the result of water being a *levelling solvent.* (See Chapter 7.) HF is relatively weak owing to the high

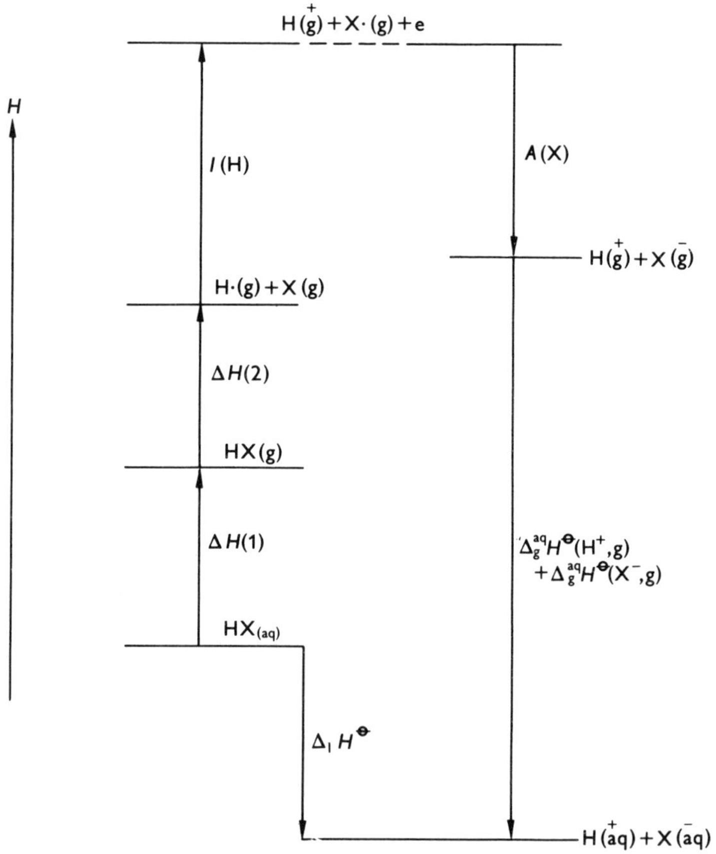

FIG. 3.11. Enthalpy diagram for the ionization of hydrohalic acids.

H–F bond energy, and the existence of associated species in solution. In an acidic solvent, such as glacial acetic acid, the order of acid strength becomes:

$$\mathrm{HI} > \mathrm{HBr} > \mathrm{HCl} \gg \mathrm{HF}$$

This is the reverse to what might be expected. Figure 3.11 summarizes the thermochemical steps in the ionization of HX.

The enthalpy of ionization is given by

$$\Delta_1 H^{\ominus} = A(\mathrm{X}) + \Delta_{\mathrm{g}}^{\mathrm{aq}} H^{\ominus}(\mathrm{H}^+, \mathrm{g}) + \Delta_{\mathrm{g}}^{\mathrm{aq}} H^{\ominus}(\mathrm{X}^-, \mathrm{g}) + I(\mathrm{H}) + \Delta H(2) + \Delta H(1)$$

For a series, the second and fourth terms are constant. What are the remaining important terms in this expression? If we knew this, we could be sure that there would be a similar variation in $\Delta_I G^{\ominus}$ (and hence in the dissociation constants) since the entropy of ionization is expected to be similar throughout the series.

Figure 3.12 shows the variation in each of the important terms.

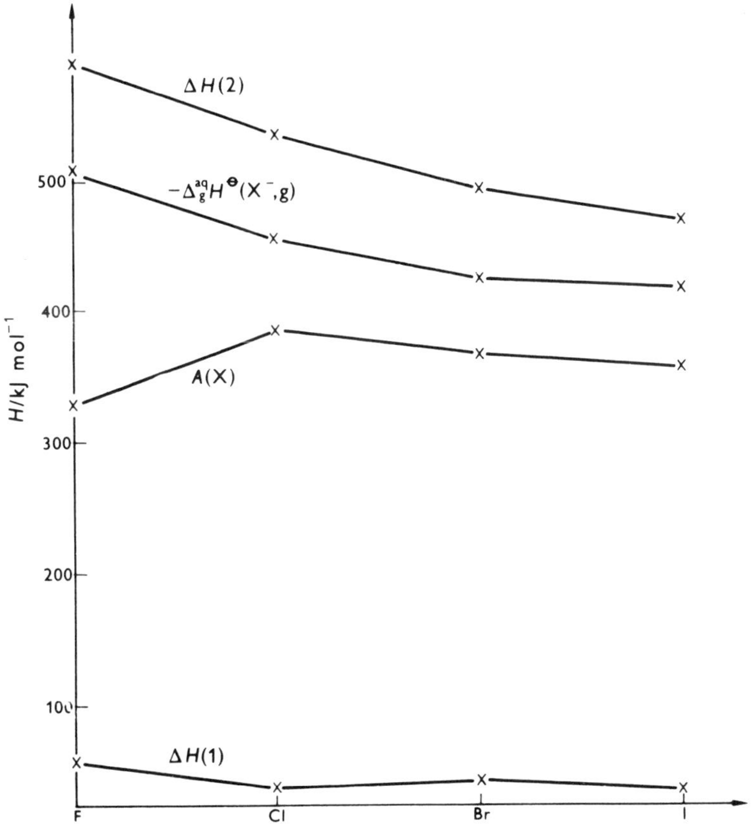

FIG. 3.12. Dependence of enthalpy terms on the type of halogen.

$\Delta H(1)$ is rather unimportant. The low value of the electron affinity, A(F), and the high bond energy of H–F(g) are responsible for the more positive value of $\Delta_I H$ for HF. For the rest of the series, the decreasing value of D(X–H), i.e. $\Delta H(2)$, is the reason for the increasing ease of ionization of HX(aq) but it is clear from the figure that the variation in each term becomes closely parallel and little discernible variation in acidity in aqueous solution is expected. The order of $\Delta_I H$ is F > Cl ≈ Br ≈ I. In a weakly solvating solvent (where $\Delta_g H$ (X^-, g) and $\Delta H(1)$ are much smaller) the variation in $\Delta H(2)$ is much more important. Since this decreases sharply down the group, the order of $\Delta_I H$ becomes

$$F > Cl > Br > I$$

This is, in fact, the order to be expected in solvents such as glacial acetic acid.

3.6.8 Noble Gas Compounds [10]

The most stable compounds of this type are those of xenon, and the majority of these are oxides, fluorides, and related compounds. The formation of these gaseous noble gas compounds, NB_y, where N is a noble gas atom, is represented energetically in Fig. 3.13. For the reaction to be spontaneous,

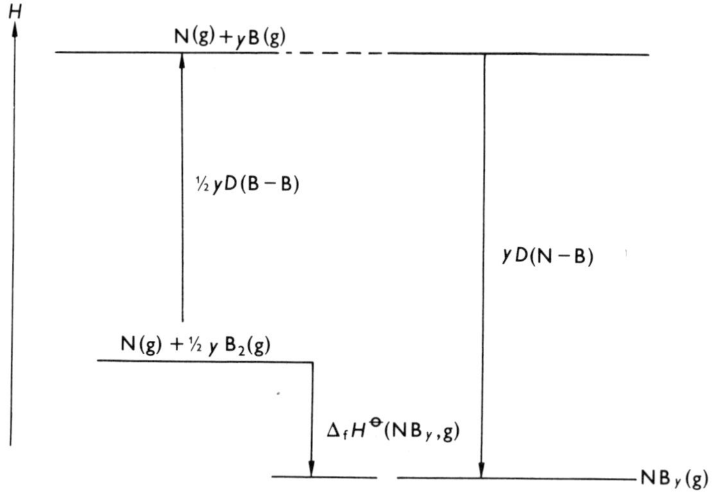

FIG. 3.13. Enthalpy diagram for the formation of noble gas compounds.

D(N–B) must exceed 1/2 D(B–B) since ΔS is expected to be negative in such a reaction. D(N–B) must, therefore, exceed 77, 122, and 249 kJ mol^{-1} for fluorides, chlorides, and oxides respectively.

Thermodynamic data for three xenon compounds are given in Table 3.10.

Table 3.10. Thermochemical Properties of some Xenon Compounds

Compound	$\Delta_f H(s)$/kJ mol^{-1}	$\Delta_s^g H$/kJ mol^{-1}	D(Xe–B)
XeF_4	−294	63	134
XeF_6	−370	62·6	128
XeO_3	399	—	<118

D(Xe–O) is an upper limit and the condition $D(N\text{–}B) > 1/2D(B\text{–}B)$ is clearly not met for the very unstable compound XeO_3.

We can use these data to predict the stability with respect to the elements of, for example, a series of oxyfluorides. Thus for gaseous $XeOF_4$, we have

$$\Delta_f H(XeOF_4) = -4D(Xe\text{–}F) - D(Xe\text{–}O) + 2D(F\text{–}F) + 1/2D(O\text{–}O)$$

For gaseous XeO_2F_2,

$$\Delta_f H = -2D(Xe\text{–}F) - 2D(Xe\text{–}O) + D(F\text{–}F) + D(O\text{–}O)$$

From the values of D(Xe–B), given above, we calculate the approximate values

$$\Delta_f H(XeOF_4, g) = -101 \text{ kJ mol}^{-1}$$
$$\Delta_f H(XeO_2F_2, g) = +147 \text{ kJ mol}^{-1}$$

And these values account for the instability of XeO_2F_2. Extrapolation of the M–Cl bond energy in a periodic series yields an estimate of the Xe–Cl bond energy:

Compound	$SnCl_2$	$SbCl_3$	$TeCl_4$	ICl_3	($XeCl_4$)
D(M–Cl)/kJ mol^{-1}	399	315	231	192	~21

It is not surprising, therefore, that $XeCl_4$ has not been prepared since $\frac{1}{2}D(Cl\text{–}Cl) \gg D(Xe\text{–}Cl)$. A similar extrapolation for AsF_3, SeF_4, and BrF_5 yields $D(Kr\text{–}F) = 80$ kJ mol^{-1} in KrF_6. This again precludes the possibility of KrF_6 being a stable compound. KrF_2 appears to have been prepared and this is acceptable since the decreased steric crowding must lead to a much larger D(Kr–F).

A simple molecular orbital treatment [11] of the noble gas fluorides considers linear units of $X_1\text{–}Xe\text{–}X_2$. An LCAO treatment predicts that the overlap of a xenon p-orbital with X p-orbitals gives rise to three molecular orbitals per $X_1\text{–}Xe\text{–}X_2$ unit. The orbitals are:

$$\psi_1 = a(X_1)\,\psi(X_1) + b(Xe)\,\psi(Xe) + a(X_2)\,\psi(X_2)$$
$$\psi_2 = c(X_1)\,\psi(X_1) - c(X_2)\,\psi(X_2)$$
$$\psi_3 = a(X_1)\,\psi(X_1) - b(Xe)\,\psi(Xe) + a(X_2)\,\psi(X_2)$$

(where *a, b,* and *c* are normalized coefficients of the appropriate atomic orbitals). The energy level diagram is then as in Fig. 3.14.

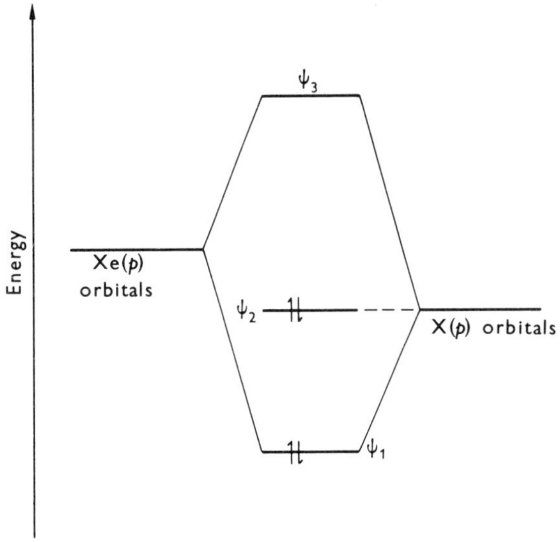

FIG. 3.14. Energy level diagram for linear X_1–Xe–X_1 units.

ψ_2 is non-bonding and the coefficient of ψ(Xe) is zero. Since each *p*-orbital of xenon contains two electrons, ψ_2 must contain two electrons. The charge densities in ψ_2 reside entirely on the X atoms and it is unreasonable to see how this would occur except with the more electronegative elements. Whilst this theory is clearly an approximation, it accounts nicely for the XeF_n (n = even) formulae.

REFERENCES

The Inert Pair Effect

[1] Drago, R. S., *J. phys. Chem.*, **62**, 353 (1958).

[2] Dasent, W. E., *Non-Existent Compounds*, p. 77, Marcel Dekker, New York (1965).

Lattice Energies

[3] Harvey, K. B., and Porter, G. B., *Introduction to Physical Inorganic Chemistry*, Addison–Wesley, Reading, Mass. (1963).

[4] Waddington, T. C., *Adv. inorg. Chem. and Radiochem.*, **1**, 158 (1959).

[5] Quane, D., *J. chem. Ed.*, **47**, 396 (1970).

Stabilization of Oxidation States

[6] Orgel, L. E., *An Introduction to Transition Metal Chemistry*, 2nd edn., Methuen, London (1966).

Addition Compounds
[7] Brown, H. C., and Holmes, R. R., *J. amer. chem. Soc.*, **78**, 2173 (1956).

Hydrides
[8] Lipscomb, W. N., *Adv. inorg. Chem. and Radiochem.*, **1**, 117 (1959).
[9] Aylett, B. J., *Adv. inorg. Chem. and Radiochem.*, **11**, 259 (1968).

Noble Gas Compounds
[10] Bartlett, N., *Endeavour*, **88**, 3(1963).
[11] Rundle, R. E., *J. amer. chem. Soc.*. **85**, 112 (1967).

Chapter 4

Transition Elements

4.1 Introduction

Transition elements are those with incomplete *d* or *f*-electron shells in either the elemental state or in the commonly occurring oxidation states. Table 4.1 shows the transition elements in relation to the whole periodic table, and also includes the valence electronic structures of the elements (the electrons outside the structure of the preceding inert gas).

On forming ions, the first electrons to be lost are those from the *s*-orbitals, followed by the number of *d*-electrons needed to produce the particular ion. In the first transition series, for example, the M(II) ions have no 4*s*-electrons but the number of *d*-electrons increases uniformly from Sc(II) with $3d^1$ to Cu(II) with $3d^9$.

On the above definition over half the elements of the periodic table belong to the transition groups. It is often convenient to include the elements immediately adjoining the various transition series, for example calcium and zinc with the first series, in order to provide a base from which to discuss the thermodynamic properties of the transition elements.

In this chapter, after a discussion of the theories of bonding in transition element compounds, we shall consider the thermodynamics of species present firstly in aqueous solution and secondly in the solid and gaseous states. Finally we shall mention briefly the calculation of coordinate bond dissociation energies from several theoretical models before going on to a discussion of the *f*-series transition elements, the lanthanides and actinides, in Section 4.6. Some aspects of metal-carbon bonding in the transition series is covered in Chapter 5.

4.2 Bonding Theories

4.2.1 Introduction

We shall be concerned mainly, in our discussion of bonding, with complexes (or coordination compounds) of transition elements. The simplest of these may be defined as compounds in which there is a central metal atom or cation surrounded by charged or neutral electron-donor atoms or molecules called

Table 4.1. Electronic Structures of Neutral Transition Elements†

Series	Group		I A	II A	III A	IV A	V A	VI A	VII A	VIII			I B	II B	III B	IV B	V B	VI B	VII B	O
			1 H																	2 He
			3 Li	4 Be											5 B	6 C	7 N	8 O	9 F	10 Ne
			11 Na	12 Mg				Transition elements							13 Al	14 Si	15 P	16 S	17 Cl	18 Ar
1st			19 K	20 Ca	21 Sc	22 Ti	23 V	24 Cr	25 Mn	26 Fe	27 Co	28 Ni	29 Cu	30 Zn	31 Ga	32 Ge	33 As	34 Se	35 Br	36 Kr
		3*d*			1	2	3	5	5	6	7	8	10	10	10	10	10	10	10	10
	[Ar] +	4*s*	1	2	2	2	2	1	2	2	2	2	1	2	2	2	2	2	2	2
		4*p*													1	2	3	4	5	6
2nd			37 Rb	38 Sr	39 Y	40 Zr	41 Nb	42 Mo	43 Tc	44 Ru	45 Rh	46 Pd	47 Ag	48 Cd	49 In	50 Sn	51 Sb	52 Te	53 I	54 Xe
		4*d*			1	2	4	5	6	7	8	10	10	10	10	10	10	10	10	10
	[Kr] +	5*s*	1	2	2	2	1	1	1	1	1	0	1	2	2	2	2	2	2	2
		5*p*													1	2	3	4	5	6
3rd			55 Cs	56 Ba	57 La*	72 Hf	73 Ta	74 W	75 Re	76 Os	77 Ir	78 Pt	79 Au	80 Hg	81 Tl	82 Pb	83 Bi	84 Po	85 At	86 Rn
		5*d*			1	2	3	4	5	6	7	9	10	10	10	10	10	10	10	10
	[Xe] +	6*s*	1	2	2	2	2	2	2	2	2	1	1	2	2	2	2	2	2	2
		6*p*													1	2	3	4	5	6
4th			87 Fr	88 Ra	89 Ac**															
	[Rn] +	6*d*			1															
		7*s*	1	2	2															

*Lanthanides	58 Ce	59 Pr	60 Nd	61 Pm	62 Sm	63 Eu	64 Gd	65 Tb	66 Dy	67 Ho	68 Er	69 Tm	70 Yb	71 Lu

**Actinides	90 Th	91 Pa	92 U	93 Np	94 Pu	95 Am	96 Cm	97 Bk	98 Cf	99 Es	100 Fm	101 Md	102 No	103 Lw

†The lonthanides and actinides are discussed in 4.6.

ligands, for example $[Co^{III}(NH_3)_6]Cl_3$; the complex is enclosed in square brackets.

Interest in the thermodynamic properties of transition element compounds has centred round two problems.

(1) The first is the problem of bonding type and the calculation of the total bond energy from some theoretical model. This is not a problem peculiar to the transition series, and the important types of bonding and bonding theories have been reviewed in Chapter 1.

For compounds with predominantly ionic bonding (hard-hard interactions) the total coordinate bond dissociation energy can be calculated with a reasonable degree of accuracy from simple electrostatic theory provided that sufficient bond length and other data are available. For softer interactions the polarizability of the ions must be allowed for. In some cases, the molecular orbital theory, with certain simplifying assumptions, allows the calculation of the total coordinate bond dissociation energy. Examples of these calculations and their limitations are given later in this chapter (Section 4.4.2).

(2) The second problem concerns the energies of the inner *d*-orbitals, the varying electronic occupancy of which gives rise to thermodynamic effects superimposed on the overall bond energy discussed above.

Three theories have been developed to tackle this second problem. Of these the valence bond theory [1] is now rarely used. The remaining two, the crystal field theory (CFT) and the molecular orbital (MO) theory have both been used to interpret thermodynamic data for transition element compounds; indeed in many cases both lead to similar predictions, but from rather different standpoints.

The CFT concentrates on the effect of an electrostatic field (from the surrounding negatively charged ions or from neutral ligands considered as dipoles) on the energies of the *d*-orbitals of the central transition element cation. This effect is superimposed on the ionic or covalent bonding already present as discussed in (1) above. The total bond energy cannot be calculated by CFT but the theory has been particularly useful in interpreting trends in *CBE*s, and relating these to spectroscopic observations. It is clear that CFT must ignore the effect of covalent bonding – a serious omission since this may be an important part of the bonding in complexes. A theory called Ligand Field Theory (or adjusted CFT) to some extent allows for covalency, but we shall not deal with this in the present text.

Molecular orbital (MO) theory is the more satisfactory approach from a theoretical viewpoint since it provides the basis for the complete calculation of metal-ligand bond energies, not merely the contributions of the *d*-electrons. A number of semi-empirical MO calculations have been performed but even the simplest of these requires a digital computer. Although we shall refer to these calculations later in this chapter, we shall restrict our discussion of MO theory to its qualitative aspects.

4.2.2 Crystal Field Theory (CFT)

The five *d*-orbitals of an isolated transition metal atom or ion are degenerate, that is, all of the same energy. This is not the case when the metal is placed in an electrostatic field provided by the ligand ions or molecules (considered as dipoles) of a complex, or indeed by the negative ions of a crystal of, for example, $CoCl_2$. We shall investigate the type and magnitude of the splitting of the *d*-orbital energy levels in the electrostatic fields produced by some common symmetries of metal complexes. The most common symmetrical arrangement of ligand molecules about a metal ion is *cubic*. Cubic symmetry is possessed by both octahedral (Fig. 4.1a) and tetrahedral (Fig. 4.1b)

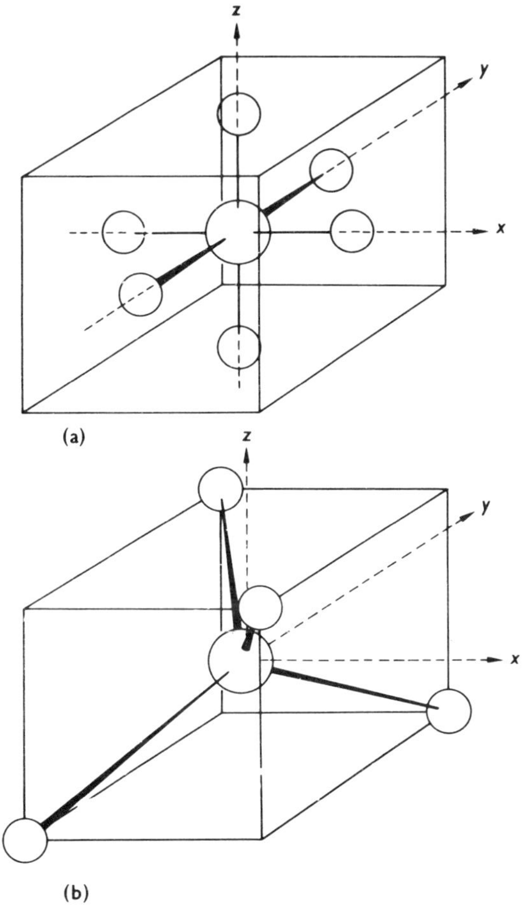

FIG. 4.1. Cubic symmetry shown by (a) octahedral and (b) tetrahedral arrangements of ligands around a metal atom or ion.

complexes. We have neglected any ligand symmetry and assumed that they are point negative charges. For an octahedral complex, Fig. 4.1(a) shows that, although the overall energy of the *d*-orbitals of the free ion will be raised by interaction with the negatively charged ligands, those orbitals that point along the defined metal-ligand axes (d_{z^2} and $d_{x^2-y^2}$) will be raised in energy to a greater extent than those that point between the axes (d_{xy}, d_{xz}, and d_{yz}).

The magnitude of the splitting depends on the strength of the electrostatic field which, in turn, depends on the electrostatic charges on the metal and ligands, and on the metal-ligand distance.

The splitting is inverted for a tetrahedral complex since the *presence* of negative charges at the cube corners (Fig. 4.1b) is equivalent to the *absence* of negative charges at the cube faces. This is the reverse of the octahedral situation.

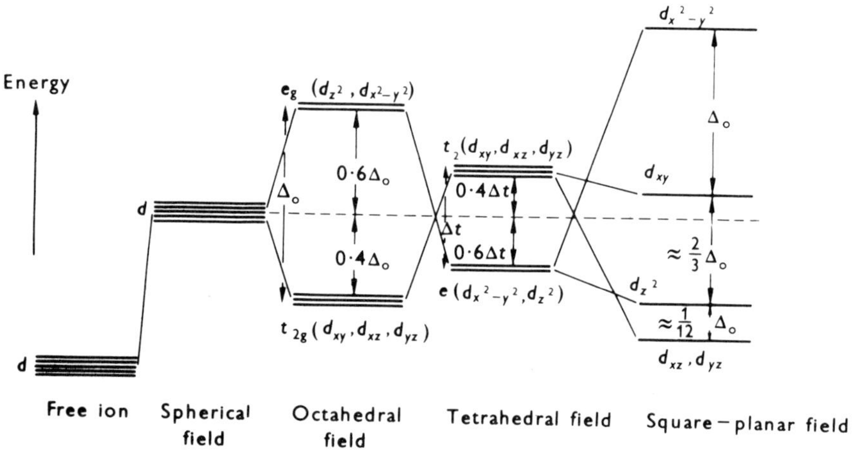

FIG. 4.2. Energy levels of *d*-orbitals in electrostatic fields of different symmetries.

Figure 4.2 shows the effect of electrostatic fields of octahedral, tetrahedral, and square-planar symmetries on the energies of the *d*-orbitals relative to their energies in a spherically-symmetrical field. The *d*-orbitals are degenerate in a spherical field but their energy is, of course, higher than in the free ion, although the CFT is not concerned with this difference. The group of orbitals d_{xy}, d_{xz}, and d_{yz} are referred to as t_{2g} orbitals, and the $d_{x^2-y^2}$ and d_{z^2} as e_g orbitals in fields of octahedral symmetry, these terms being of group-theoretical origin [2]. Other group-theory terms for tetrahedral fields are shown in Fig. 4.2. It should be noted that whereas bonding between metal and ligands reduces the overall energy of the system, the *d*-orbitals of the metal, and it is these on which the CFT concentrates, will be raised in energy compared with the free metal ion.

It is usual to say that the *d*-orbital energies are split by the electrostatic field by an amount Δ_o in an octahedral field and by an amount Δ_t in a tetrahedral field, as shown in Fig. 4.2. Along a transition series the increasing numbers of *d*-electrons will fill the orbitals beginning with those of lowest energy but with the restriction that, unless there is a very large value of Δ

(greater than that needed to force electrons of opposite spin to occupy the same orbital) there will be maximum unpairing of spins.

Depending on the number of *d*-electrons and on the symmetry of the crystal field, there will be varying amounts of crystal field stabilization energy, ΔH(CFSE), defined as the stabilization produced by placing the electrons in the split orbitals compared with placing them in degenerate orbitals in an equal field of spherical symmetry. Table 4.2 summarizes the electronic configurations and stabilization energies for the more usual 'high-spin' octahedral and tetrahedral complexes, and for the 'low-spin' situation. This last arises for ligands such as CN^- which produce fields of large Δ_o. In cyanide complexes, electrons prefer to be paired in lower energy t_{2g} orbitals before occupying the higher energy e_g orbitals. 'Low-spin' tetrahedral complexes are very rare.

Expressed in equation form the stabilizations, which contribute positively to the total coordinate bond dissociation energy, for 'high-spin' cases are

$$\Delta H(\text{CFSE}) = \Delta_o[-0{\cdot}4n(t_{2g}) + 0{\cdot}6n(e_g)] \quad \text{octahedral}$$

$$\Delta H(\text{CFSE}) = \Delta_t[-0{\cdot}6n(e) + 0{\cdot}4n(t_2)] \quad \text{tetrahedral}$$

where n is the number of electrons in the orbitals shown in Fig. 4.2.

The energy required to pair the electrons need not be taken into account in calculating stabilization energies for the 'high-spin' cases since the distribution of *d*-electrons between the *d*-orbitals will not differ from that in the free ion or an ion in a spherical field. However, in a 'low-spin' complex where some electrons are forced to pair before all the *d*-orbitals are singly occupied, the pairing energy must be taken into account. Actually it is more usual to express the situation in terms of the stabilization or 'exchange' energy K acquired by the system for each pair of electrons with parallel spins. The relative magnitude of this energy is shown in Table 4.2 for 'high-spin' and 'low-spin' octahedral complexes. Differences between the two types occur from d^4 to d^7 electronic configurations and must be taken into account in the treatment of the thermodynamics of the 'low-spin' systems. Unfortunately the actual values of the exchange energies are rarely known. Calculations [3] indicate that the value of K for $[Fe^{II}(H_2O)_6]^{2+}$, for example, is about 83 kJ mol^{-1}, and that this does not vary greatly with the ligand.

Since CFT predicts a stabilization of certain electronic configurations, it is important to know the values of the splitting parameters Δ_o and Δ_t. The theory gives no indication of these values but spectroscopic visible and ultraviolet absorption measurement of transition metal complexes can usually be analysed to provide values of Δ, some examples of which are given in Table 4.3. We are now in a position to predict from spectroscopic measurements the variation of thermodynamic stability of a series of complexes. We

Table 4.2. Crystal field stabilization energies [these count positively to *CBE* (metal-ligand)] and exchange energies for 'high-' and 'low-spin' electron arrangements in octahedral and tetrahedral electrostatic fields.

Number of *d*-electrons	Corresponding M(II) ion[a]	Octahedral field				Tetrahedral field	
		High-spin		Low-spin		High-spin	
		ΔH(CFSE)	Exchange energy	ΔH(CFSE)	Exchange energy	ΔH(CFSE)	Exchange energy
0	Ca	$0\ \Delta_o$	$0K$	$0\ \Delta_o$	$0K$	$0\ \Delta_t$	$0K$
1	Sc	$0{\cdot}4\ \Delta_o$	$0K$	$0{\cdot}4\ \Delta_o$	$0K$	$0{\cdot}6\ \Delta_t$	$0K$
2	Ti	$0{\cdot}8\ \Delta_o$	$1K$	$0{\cdot}8\ \Delta_o$	$1K$	$1{\cdot}2\ \Delta_t$	$1K$
3	V	$1{\cdot}2\ \Delta_o$	$3K$	$1{\cdot}2\ \Delta_o$	$3K$	$0{\cdot}8\ \Delta_t$	$3K$
4	Cr	$0{\cdot}6\ \Delta_o$	$6K$	$1{\cdot}6\ \Delta_o$	$1K$	$0{\cdot}4\ \Delta_t$	$6K$
5	Mn	$0\ \Delta_o$	$10K$	$2{\cdot}0\ \Delta_o$	$0K$	$0\ \Delta_t$	$10K$
6	Fe	$0{\cdot}4\ \Delta_o$	$6K$	$2{\cdot}4\ \Delta_o$	$0K$	$0{\cdot}6\ \Delta_t$	$6K$
7	Co	$0{\cdot}8\ \Delta_o$	$3K$	$1{\cdot}8\ \Delta_o$	$0K$	$1{\cdot}2\ \Delta_t$	$3K$
8	Ni	$1{\cdot}2\ \Delta_o$	$1K$	$1{\cdot}2\ \Delta_o$	$1K$	$0{\cdot}8\ \Delta_t$	$1K$
9	Cu	$0{\cdot}6\ \Delta_o$	$0K$	$0{\cdot}6\ \Delta_o$	$0K$	$0{\cdot}4\ \Delta_t$	$0K$
10	Zn	$0\ \Delta_o$	$0K$	$0\ \Delta_o$	$0K$	$0\ \Delta_t$	$0K$

a of the first transition series.

Table 4.3. Spectroscopically Determined Values of Crystal Field Splitting Parameters, Δ/kJ mol^{-1}, for Transition Metal Complexes. Italics indicate 'low-spin' complexes. Data are mainly from Orgel [4] and Figgis [5]. Literature values vary somewhat.

Transition metal ion	Electronic structure	Octahedral					Tetrahedral
		$6H_2O$	$6NH_3$	3en	2dien	$6CN^-$	$4Cl^-$
		Δ_o	Δ_o	Δ_o	Δ_o	Δ_o	Δ_t
Ti(II)	$3d^2$	263					
Ti(III)	$3d^1$	243	203				
V (II)	$3d^3$	147				*166*[a]	
V(III)	$3d^2$	223	215				
Cr(II)	$3d^4$	169		215		*305*[a]	
Cr(III)	$3d^3$	203	263	262		315	
Mn(II)	$3d^5$	102		109		*285*[a]	
Mn(III)	$3d^4$	239					
Fe(II)	$3d^6$	120		136	136	*395*	48
Fe(III)	$3d^5$	167					
Co(II)	$3d^7$	111	121	132	132	*407*	37
Co(III)	$3d^6$	*218*	*287*	*279*			
Ni(II)	$3d^8$	106	129	139	139		43
Cu(II)	$3d^9$	144	181	196			58
Rh(III)	$4d^6$	*323*	*407*	*412*			
Ir(III)	$5d^6$			*493*			

a determined calorimetrically.

shall see later that good agreement exists between the two series of measurements, an agreement which is strong support for the CFT.

Values of Δ_o are larger for trivalent ions compared with divalent ions of the first series, and there is a further increase of 30–50 per cent in Δ_o from first to second and from second to third transition series. These values are sufficiently large to make a significant contribution to the overall bond energy of transition metal compounds. Useful approximations are that

$$\Delta_t = \frac{4}{9}\Delta_o$$

and that for a complex with mixed ligands, Δ is the weighted average of the Δ values for each separate ligand.

Note that, in the absence of crystal field stabilization, the theory gives no indication of the variation of bond energy along a transition series of complexes. It is reasonable to assume, from electrostatic considerations related to the changing ionic radii, that the variation will be a smooth, though not necessarily linear one.

Many properties of transition metal compounds, such as magnetic and structural characteristics, are influenced by the electrostatic field. In this chapter we shall be concerned only with the thermodynamic aspects.

Tetragonal Distortion

If a metal complex has an orbitally degenerate ground state, the complex distorts so that the degeneracy is removed. This is the Jahn–Teller theorem. This distortion causes further splitting of the *d*-orbital energies with a consequent small increase in ΔH(CFSE).

Octahedral complexes commonly undergo trigonal or tetragonal distortions in such circumstances. The tetragonal distortion (axial lengthening),

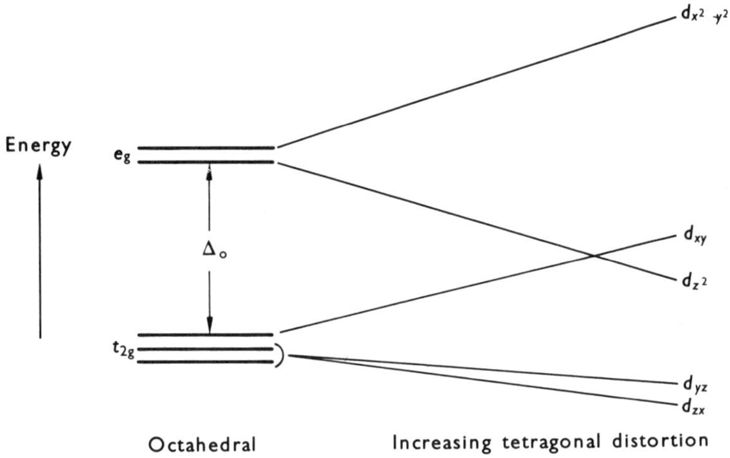

FIG. 4.3 Further splitting of the *d*-orbital energy levels on tetragonal distortion of an octahedral complex. (From Reference [6].)

shown in Fig. 4.3, is the more common and has been observed for

$t_{2g}^3 e_g^1$ e.g. 'high-spin' Cr(II) and Mn(III)

$t_{2g}^6 e_g^1$ e.g. 'low-spin' Co(II) and Ni(III)

$t_{2g}^6 e_g^3$ e.g. 'high-spin' Cu(II)

Examples of Jahn–Teller stabilization for the most common case, Cu(II), are mentioned in the following sections, The limiting case of tetragonal distortion is a square-planar situation when the axial ligands are completely removed. The splitting of the *d*-orbitals is then of the form shown earlier in Fig. 4.2.

4.2.3 Molecular Orbital (MO) Theory

The CFT has a number of unsatisfactory aspects notably its neglect, by concentrating on electrostatic influences on the *d*-electrons, of any possibility of covalent or part-covalent bonding. Spectroscopic, nuclear magnetic resonance, and electron spin resonance experiments [7] indicate that covalent contributions to the bonding of transition metal complexes are far from negligible.

The MO theory (Section 1.3) considers that the overlap of atomic orbitals to produce molecular orbitals occurs whenever symmetry permits. The amount of overlap is related to the amount of covalency of the bonding. The process of deciding which orbitals are of suitable symmetry for overlap is conveniently dealt with by group theory [2], outside the scope of the present text, but many of the symbols currently used in the discussion of bonding in transition metal compounds are of group theoretical origin.

In MO theory it is possible to consider all the nine valence shell orbitals of *s, p,* and *d* type in the manner now briefly summarized for the case of an octahedral complex. Six of the valence shell orbitals on the transition metal atom or ion, the e_g (d_{z^2} and $d_{x^2-y^2}$), the $a_{1g}(s)$, and the t_{1u} (p_x, p_y, and p_z) lie along the six octahedral directions and are suitable for σ-bonding by overlap with appropriate ligand orbitals. The remaining metal orbitals, the t_{2g} (d_{xy}, d_{xz}, and d_{yz}) lie between the octahedral directions and are suitable only for π-bonding.

The procedure is now to find linear combinations of σ-orbitals on the ligands to produce orbitals of the same symmetry properties (e_g, a_{1g}, and t_{1u}) as the metal orbitals in question. Overlap of the metal and ligand orbitals produces bonding and antibonding (*) molecular orbitals as described in Section 1.3. If there are π-orbitals on the ligand these too may be combined into appropriate symmetry (t_{2g}) orbitals for combination with metal t_{2g} orbitals.

Figure 4.4 shows an energy diagram for an octahedral complex formed from a first transition series metal. The metal 3*d*, 4*s*, and 4*p* orbitals, also described by the group symbols e_g and t_{2g}, a_{1g}, and t_{1u}, respectively, are combined with ligand orbitals of suitable symmetry to produce the bonding molecular orbitals of lower energies and antibonding (*) orbitals of higher energies than the original orbitals. Note that of the five 3*d* orbitals, only those two of symmetry e_g have combined with ligand orbitals of equivalent symmetry. The remaining three 3*d* orbitals labelled t_{2g} have no corresponding ligand t_{2g} orbitals with which to combine to form σ-bonds, and are shown as being unchanged in energy. Of course if the ligand has suitable t_{2g} orbitals (many ligands do not), in addition to the σ-type, then combination of these would lead to π-bond formation with a consequent change in the energy of the t_{2g} orbitals. This is discussed more fully below.

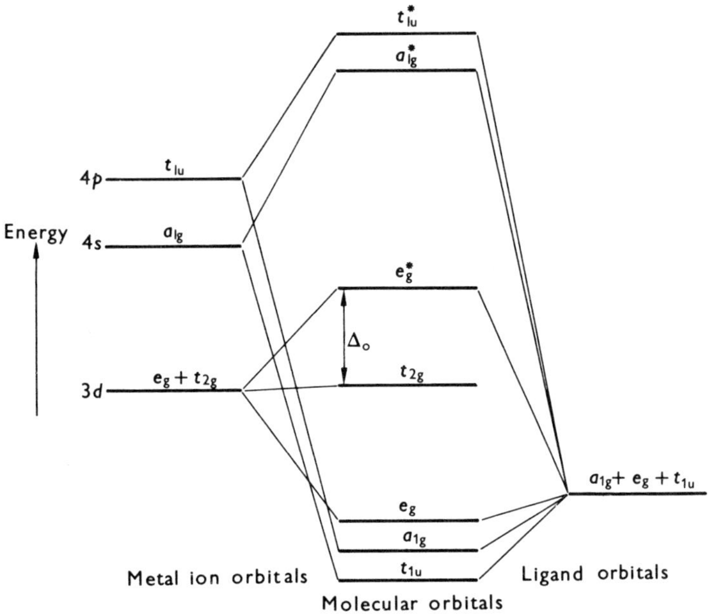

FIG. 4.4. Molecular orbital energy level diagram octahedral complex of a first transition series metal. (From Reference [6].)

At the centre of Fig. 4.4, the antibonding molecular orbital, e_g^*, is of higher energy by Δ_o than the unchanged t_{2g} orbitals. The MO theory thus predicts, in agreement with the CFT but from a completely different standpoint, a splitting of the energy levels of the five degenerate 3*d* metal orbitals into two groups with the d_{z^2} and $d_{x^2-y^2}$ (e_g^*) at a higher energy than the d_{xy}, d_{yz}, and d_{xz} (t_{2g}) orbitals.

Figure 4.5 shows the effect of π-bonding through orbitals of t_{2g} symmetry. Case (a) refers to ligands such as amines and phosphines with empty π-orbitals (usually *d*-type) of higher energy than the metal t_{2g} orbitals. The effect of this is to increase the value of Δ_o from that expected for σ-interactions alone. Case (b) refers to ligands such as oxygen and fluorine with filled π-orbitals (usually *p*-type) of lower energy than the metal t_{2g} orbitals, for which a reduction in the value of Δ_o occurs. Predictions of the effect of ligands with both empty and filled π-orbitals, such as the halide and cyanide ions, are not easily made.

4.3 Aqueous Systems

4.3.1 Ligand Field Effects

The bonding of anions or neutral molecules to transition metal ions in aqueous solution involves the displacement of water molecules already attached to the metal ion:

$$[M(H_2O)_x]^{m+}(aq) + nA^{a-}(aq) = [M(H_2O)_{x-n}(A)_n]^{m+na-}(aq) + nH_2O(aq) \tag{4.1}$$

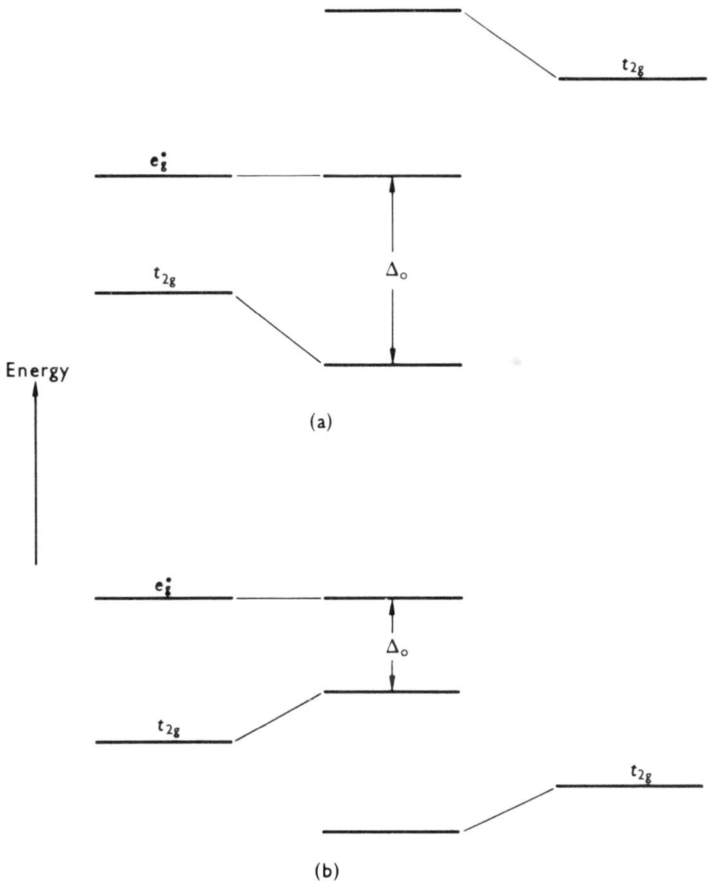

FIG. 4.5. The effect of π-bonding on the magnitude of Δ_o in a transition metal octahedral complex (a) empty ligand π-orbitals of higher energy than metal t_{2g} orbitals, (b) filled ligand π-orbitals of lower energy than metal t_{2g} orbitals. (From Reference [6].)

This was mentioned in Section 2.3.3 where it was also noted that replacement of H_2O by A occurs in steps, with each of which is associated an equilibrium constant (or stability constant) and a corresponding change in thermodynamic function. Whereas simple molecules like H_2O and NH_3 and ions such as the halides each occupy only one coordination site (unidentate ligands) on the metal ion, others may have several sites for bonding within one molecule

Table 4.4. Formulae and Abbreviations of Some Common Ligands

Name	Formula	Number of coordinating positions	Abbreviation
propionate	$CH_3CH_2COO^-$	1	prop
glycolate	$CH_2(OH)COO^-$	2	glycol
2, 2′-bipyridyl	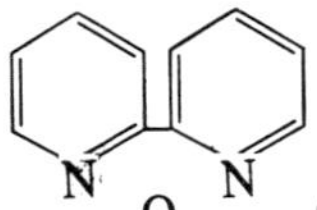	2	bipy
acetylacetonate	$CH_3-C(=O)-CH=C(O^-)-CH_3$	2	acac
ethylenediamine	$H_2NCH_2CH_2NH_2$	2	en
diethylenetriamine	$H_2NCH_2CH_2NHCH_2CH_2NH_2$	3	dien
thiourea	$S=C(NH_2)_2$	2N, 1S	tu
nitrilotriacetate	$(CH_2COO^-)_3N$	4	nta
ethylenediaminetetraacetate	$(^-OOCCH_2)_2NCH_2CH_2N(CH_2COO^-)_2$	6	edta
diethylenetriaminepentaacetate	$(^-OOCCH_2)_2NCH_2CH_2N(CH_2COO^-)CH_2CH_2N(CH_2COO^-)_2$	8	dtpa

(multidentate ligands). A few such multidentate ligands and their commonly used abbreviations are listed in Table 4.4.

In this chapter we shall express the data in terms of the reverse of the formation reaction (4.1) in order to show the parallel between this type of metal-ligand bond breaking process for aqueous species and the gas-phase *CBE* values to be discussed later in the chapter. The enthalpy change for the replacement of ligand A by H_2O in an aqueous complex is, therefore, related to the difference in the coordinate bond dissociation energy between the M–A bonds broken and the M–H_2O bonds formed, where the bond energies refer to the aqueous species. When ligand field effects operate, both types of bond will be stabilized. The net effect on the value of ΔH should, on the basis of the theories outlined in Section 4.2, be related to the difference between the values of the ligand field splitting parameter, Δ, when the overall geometry of the species does not change during the course of the reaction. For a field of octahedral symmetry the total effect on ΔH is related to the numbers of electrons, n, in the t_{2g} and e_g orbitals (determined by the atomic number and oxidation state) and to the two values of Δ by

$$\Delta H(\text{CFSE}) = [0{\cdot}4n(t_{2g}) - 0{\cdot}6n(e_g)]\ [\Delta(6\text{A}) - \Delta(6\text{H}_2\text{O})] \qquad (4.2)$$

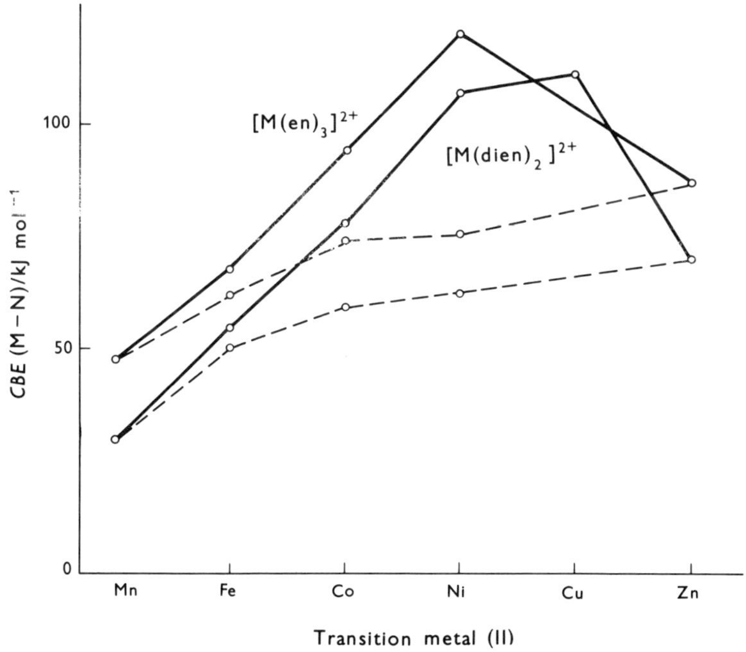

FIG. 4.6. Experimental *CBE* values for aqueous tris(ethylenediamine) and bis(diethylenetriamine) complexes [8]. Also shown (. . . o . . .) are values corrected for crystal field stabilization.

The small number of cases where sufficient experimental data are available to carry out the calculation, corroborate the above formula. Figure 4.6 shows a characteristic 'humped' graph of calorimetrically measured enthalpy changes for the *removal* of amine ligands, and their replacement by water, from some octahedral complexes of the first transition series. The complexes represented in Fig. 4.6 are $[M^{II}(en)_3]^{2+}$ and $[M^{II}(dien)_2]^{2+}$ with the bidentate ligand ethylenediamine (en) and the tridentate ligand diethylenetriamine (dien) respectively. Replacement of these ligands from the complexes requires six water molecules in each case. Values of Δ from Table 4.3 are independent of the calorimetric measurements and may be used with equation (4.2) to calculate the change in ΔH(CFSE) for the replacement of the amines by water. If these calculated stabilizations are subtracted from the measured bond energies we find that, in satisfactory agreement with CFT and MO theory, the points (Fig. 4.6) lie close to a smooth curve passing through the measured values for Mn(II) and Zn(II) for which no stabilization exists.

It is noteworthy that the value for $[Cu^{II}(dien)_2]^{2+}$ (no value is available for the Cu–ethylenediamine complex) does not agree well with the above crystal field scheme, the corrected value lying above the expected curve. The increased stability of this species may be due to Jahn–Teller distortion as discussed in Section 4.2.3. It is particularly important for Cu(II) for which the electronic configuration is $t_{2g}^6\, e_g^3$.

4.3.2 Relative Stability along the First Transition Series

The aim of this section is to correlate thermodynamic data on aqueous complexes with the aid of the hard and soft classification of acids and bases discussed in Section 1.6.

The transition metal ions of the *d*-series range from the hard Sc(III) ion to the soft ions (acids) such as Cd(II) and Hg(II), although these last two ions lie strictly outside the definition of transition metals as explained in Section 4.1.

The halide and pseudohalide anions form a convenient series of bases, monodentate and with a single negative charge, increasing in softness in the order [8] $F^- < Cl^- < Br^- < I^- < N_3^- < SCN^- < CN^-$. Along this series of bases the polarizability becomes greater, resulting in an increasing tendency towards covalent bond formation and a decreasing tendency towards ionic bond formation. Comparisons will be based on metal-ligand bond-breaking reactions in which the ligand is replaced by a water molecule.

$$[MA]^{(x-1)}(aq) = M^{x+}(aq) + A^-(aq)$$

Only 1:1 complexes are considered since the most complete data are available for these. (Note that most tabulations of thermodynamic data for aqueous complexes give the reverse, or formation, reaction.)

For reactions of the above type involving the series of halide and pseudohalide ligands, the variation in free energy change, ΔG, and the

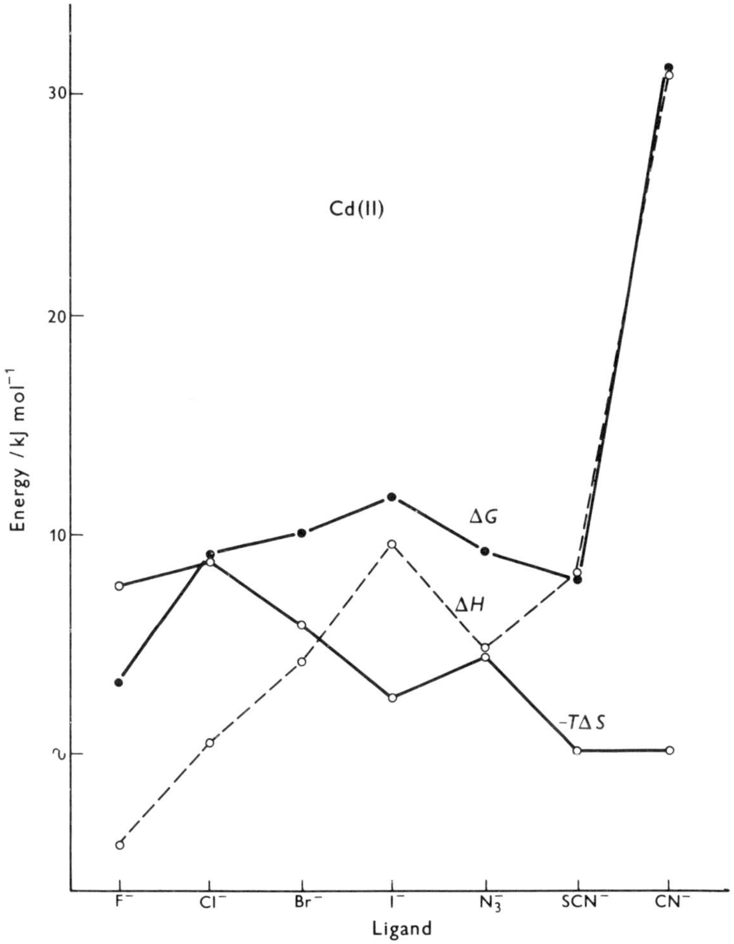

FIG. 4.7 Values [8] of ΔG, $\Delta H(= CBE)$, and $-T\Delta S$ for the replacement of the ligand by water in aqueous $[Cd^{II}(A)]^+$ complexes at 293 K in 3M $NaClO_4$.

component enthalpy (related to changes in coordinate bond dissociation energy), ΔH, and entropy, $-T\Delta S$, terms is shown in Fig. 4.7 and 4.8, respectively, for the soft ions Cd(II) and Hg(II). The thermodynamic stability of a complex $[MA]^{(x-1}$ (aq) is indicated by the positive magnitude of ΔG.

The stability increases in the order expected from F^- to CN^-, emphasizing the preference of soft acids for soft bases. The low stability of the fluoride complex shows the small attraction of this hard base for the soft Cd(II) and

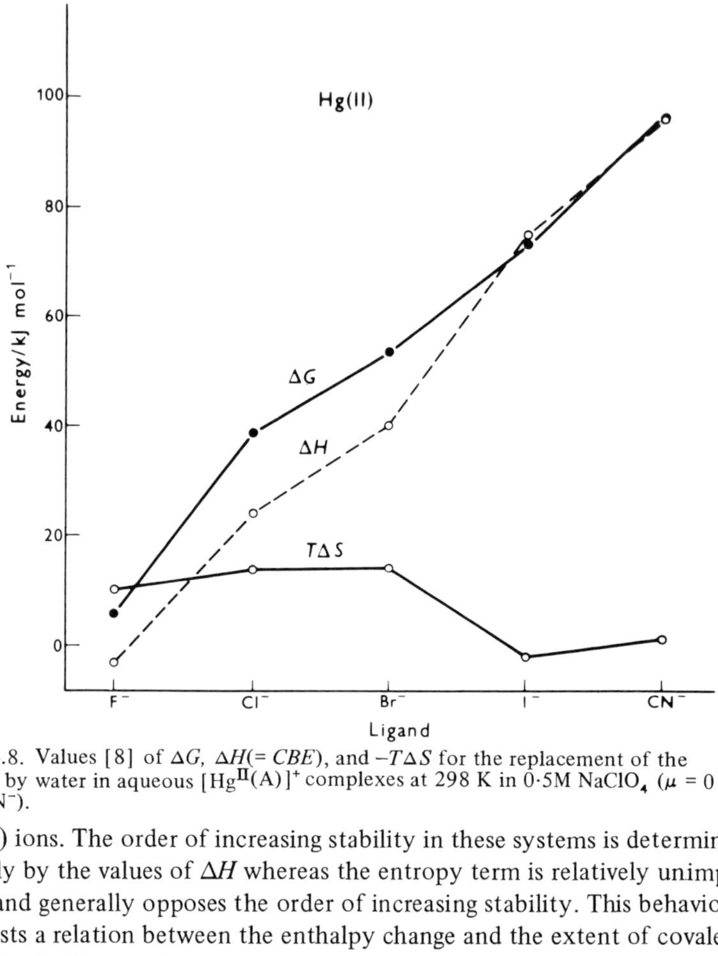

FIG. 4.8. Values [8] of ΔG, $\Delta H(= CBE)$, and $-T\Delta S$ for the replacement of the ligand by water in aqueous $[Hg^{II}(A)]^{+}$ complexes at 298 K in 0·5M $NaClO_4$ ($\mu = 0$ for CN^-).

Hg(II) ions. The order of increasing stability in these systems is determined mainly by the values of ΔH whereas the entropy term is relatively unimportant and generally opposes the order of increasing stability. This behaviour suggests a relation between the enthalpy change and the extent of covalent bonding in the complexes.

Data are also available for the hard ions Fe(III) and Sc(III), shown respectively in Figs 4.9 and 4.10. Here the order of stability is, as expected, the opposite to that observed for the soft ions Cd(II) and Hg(II) and, moreover, this order of stability is determined mainly by the entropy term. The enthalpy changes in many cases oppose the formation of complexes, particularly for the harder anions. These trends are in agreement with the expectation that hard bases prefer to bond to hard acids.

A noticeable general trend is the increasing importance of the enthalpy term, at the expense of the entropy term, to the overall stability of the complex, as the softness of both acid and base are increased. The decreasing importance of $-T\Delta S$ may be related to the smaller interaction of soft, more

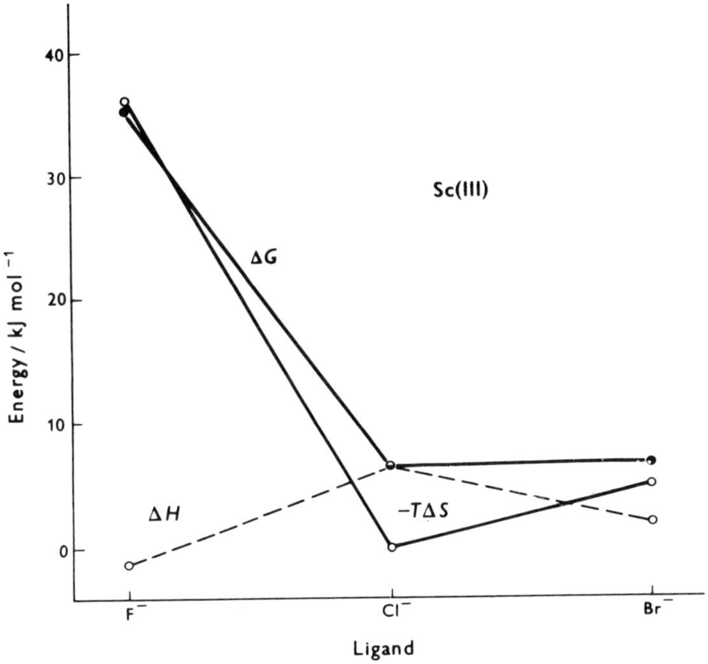

FIG. 4.9. Values [8] of ΔG, $\Delta H (= CBE)$, and $-T\Delta S$ for the replacement of ligand by water in aqueous $[Sc^{III}(A)]^{2+}$ complexes at 298 K in 0·5M $NaClO_4$.

polarizable anions with the solvent water molecules, with consequently less release of water molecules on combination of the anion with the cation.

4.3.3 Relative Stability down a Group

The great variety of compounds formed by the transition elements has so far eluded any quantitative rationalization of their relative stability. The situation is further hindered by the scarcity of thermodynamic data on the second and third transition series.

Amongst aqueous species, it is generally, but by no means invariably, found that the higher oxidation states become increasingly stable down a particular group. The data in Table 4.5, for example, show an increase in stability (less positive ΔG) of the trivalent aqueous ions relative to the divalent species from iron to ruthenium as measured by the oxidation reaction

$$M(II)(aq) = M(III)(aq)$$

Considerable and unpredictable changes in the thermodynamic functions may occur on replacing the H_2O of the aquo complexes by other ligands,

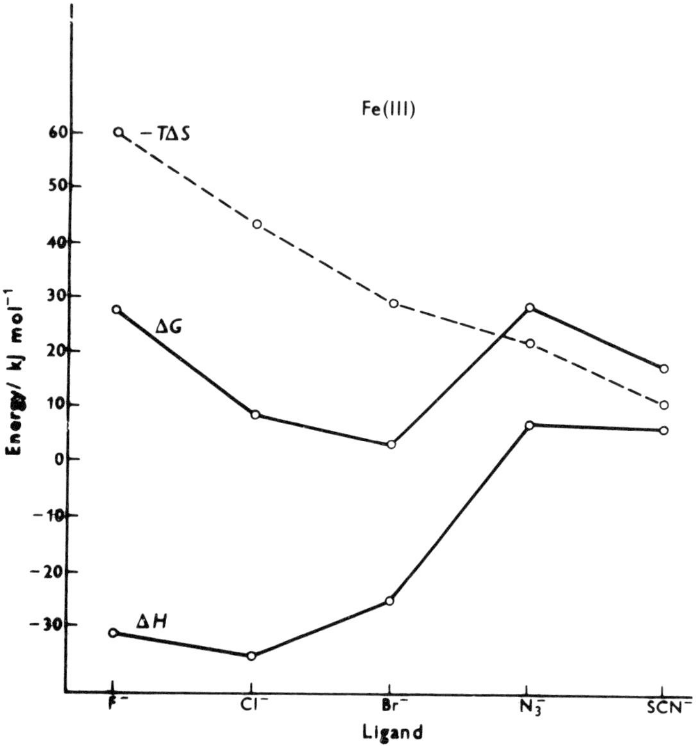

FIG. 4.10. Values [8] of ΔG, $\Delta H(= CBE)$, and $-T\Delta S$ for the replacement of ligand of water in aqueous $[Fe^{III}(A)]^{2+}$ complexes at 293 K and zero ionic strength.

still in aqueous solution. For example, the relative stability of the aqueous iron (II) ion relative to the iron (III) ion is considerably reduced by coordination of cyanide, but increased by coordination of bipyridyl, as shown by the data in Table 4.5. Strangely the Ru(II) species become relatively more stable than the Ru(III) species upon coordination both by cyanide and by bipyridyl. It is interesting to note that for iron, for which enthalpy and entropy data are available, the entropy change is of overriding importance when comparing the aquo and cyanide complexes, since the enthalpy change alone would relatively stabilize the Fe(II) species. Amongst the factors influencing ΔH are

(1) the very different charges on the hexacyanide and the hexaaquo complexes, and

(2) the differences in crystal field stabilization, complicated by the 'low-spin' configuration of the hexacyanide complex and the 'high-spin' configuration of the hexaaquo complex.

Table 4.5. Thermodynamic Data [8–10] at 298 K and Zero Ionic Strength for the Oxidation of Aqueous Iron and Ruthenium Complexes According to M(II)(aq) + H^+(aq) = M(III)(aq) + 0·5H_2(g)

M(II) complex	ΔG/kJ mol^{-1}	ΔH/kJ mol^{-1}	ΔS/J K^{-1} mol^{-1}
$[Fe^{II}(H_2O)_6]^{2+}$	74·3	40·6	−113
$[Ru^{II}(H_2O)_6]^{2+}$	24·0	−42·3	−222
$[Fe^{II}(CN)_6]^{4-}$	27·2	110	277
$[Ru^{II}(CN)_6]^{4-}$	86·8		
$[Fe^{II}(bipy)_3]^{2+}$	107	137	101
$[Ru^{II}(bipy)_3]^{2+}$	132	152	67

4.4 Solid and Gaseous Systems

4.4.1 Coordinate Bond Dissociation Energies

The effects of the environment of a molecule, solid or aqueous for example, on the bond energies are difficult to allow for. For this reason it is the energies of the bonds in the gas phase, without the added complication of the environment, which are preferred in the theoretical treatment of bonding.

The method of calculating coordinate bond dissociation energies from calorimetric and spectroscopic data was outlined in Section 2.4 but is given in more detail in the cycles shown in Fig. 4.11. It is interesting to examine certain trends in values calculated in this way before looking in more detail at some theoretical calculations of bond energies (Section 4.4.3).

Table 4.6 lists mean coordinate bond dissociation energies, $\overline{CBE}$, for a number of halides, both neutral and ionic. The values refer to reactions of the type

$$[MA_n]^{(x+n-)}(g) = M^{x+}(g) + nA^-(g) \quad \overline{CBE}(M{-}A) = \Delta H/n$$

where the total enthalpy change is divided by the number of M–A bonds.

In the gas phase the bonding between ions of opposite charge would be expected to be predominantly electrostatic in character and, since none of the energy needed to break a metal-ligand bond can come from the formation of other bonds as in the replacement of a ligand by water in aqueous systems, the bond energies would be expected to be large. Moreover the absence of solvation will render the entropy term relatively small so that stability relative to the ions is $\Delta G \approx \Delta H = CBE$.

An examination of the $\overline{CBE}$ values for the first transition series dihalides shows that the values are indeed large and that all the transition metal ions,

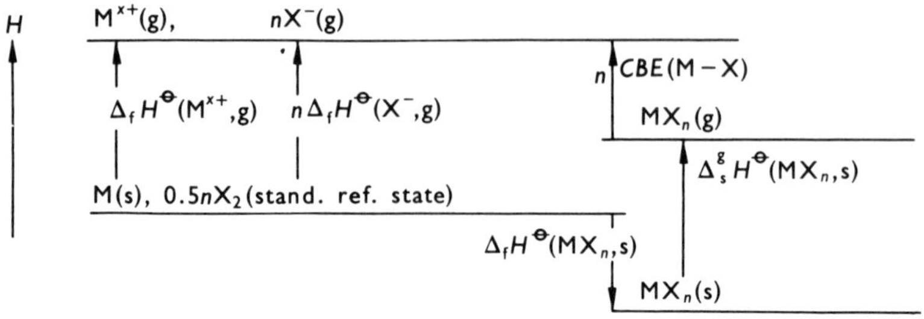

$$n\,CBE\,(\mathrm{M-X}) = -\Delta_f H^\ominus(\mathrm{MX}_n,\mathrm{s}) + \Delta_f H^\ominus(\mathrm{M}^{x+},\mathrm{g}) + n\,\Delta_f H^\ominus(\mathrm{X}^-,\mathrm{g}) - \Delta_s^g H^\ominus(\mathrm{MX,s})$$

(a)

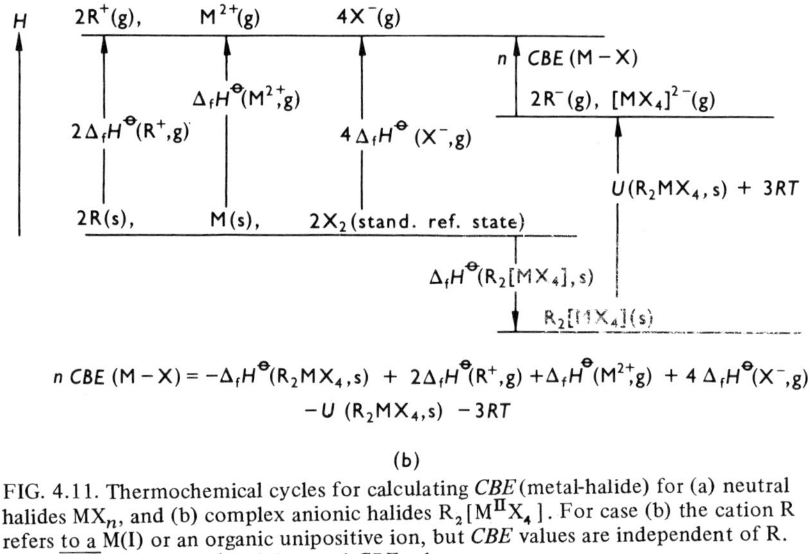

$$n\,CBE\,(\mathrm{M-X}) = -\Delta_f H^\ominus(\mathrm{R_2MX_4,s}) + 2\Delta_f H^\ominus(\mathrm{R^+,g}) + \Delta_f H^\ominus(\mathrm{M^{2+},g}) + 4\,\Delta_f H^\ominus(\mathrm{X^-,g}) - U\,(\mathrm{R_2MX_4,s}) - 3RT$$

(b)

FIG. 4.11. Thermochemical cycles for calculating *CBE*(metal-halide) for (a) neutral halides MX_n, and (b) complex anionic halides $R_2[M^{II}X_4]$. For case (b) the cation R refers to a M(I) or an organic unipositive ion, but *CBE* values are independent of R. Mean $\overline{CBE}$ values are $1/n$ of the total *CBE* values.

including the soft (in aqueous solution) cadmium and mercury ions, have halides which decrease in stability in the order expected for hard acids, $F^- > Cl^- > Br^- > I^-$. However, the rate of decrease in $\overline{CBE}$ from F^- to I^- is clearly much slower for those ions such as Cd(II) and Hg(II) compared with those considered hard in aqueous solution such as Sc(III). Thus it appears that even in the gas phase covalent bonding still contributes significantly to the bond energy between soft acids and bases.

The rate of decrease of $\overline{CBE}$ from F^- to I^- for a particular metal ion, defined as $[\overline{CBE}(F^-) - \overline{CBE}(I^-)]/\overline{CBE}(F^-)$, is therefore a measure of the softness of the ion, its value decreasing with increasing softness. Softness parameters calculated in this way have been tabulated [11] in Table 1.3.

Table 4.6. Mean Metal-halide Coordinate Bond Dissociation Energies at 298 K (Data from References [8, 11, 12].)

Metal	*d*-electron configuration	$\overline{CBE}$(M–X)/kJ mol^{-1}			
		F^-	Cl^-	Br^-	I^-
M(I)					
H	–	1538	1388	1347	1308
Cu	$3d^{10}$	831	757	749	738
Ag	$4d^{10}$	748	695	695	693
Au	$5d^{10}$	862	813	815	824
M(II)					
Ca	$3d^0$	1095	970	941	897
Sc	$3d^1$	1196	1071	1037	994
Ti	$3d^2$	1259	1139	1105	1061
V	$3d^3$	1283	1172	1143	1105
Cr	$3d^4$	1259	1158	1124	1081
Mn	$3d^5$	1240	1158	1119	1085
Fe	$3d^6$	1312	1216	1177	1143
Co	$3d^7$	1336	1230	1201	1163
Ni	$3d^8$	1375	1264	1230	1201
Cu	$3d^9$	1394	1298	1283	1250
Zn	$3d^{10}$	1389	1293	1264	1230
Sr	$4d^0$	1028	917	888	849
Cd	$4d^{10}$	1245	1182	1168	1143
Ba	$5d^0$	975	864	839	796
Hg	$5d^{10}$	1346	1283	1269	1259
Complex M(II)					
$[MnX_4]^{2-}$	$3d^5$	–	579	560	532
$[FeX_4]^{2-}$	$3d^6$	–	601	583	555
$[CoX_4]^{2-}$	$3d^7$	–	622	604	583
$[NiX_4]^{2-}$	$3d^8$	–	628	613	586
$[CuX_4]^{2-}$	$3d^9$	–	641	626	–
$[ZnX_4]^{2-}$	$3d^{10}$	–	636	619	594
$[HgX_4]^{2-}$	$5d^{10}$	–	–	644	620

Table 4.6. (*continued*)

Metal	*d*-electron configuration	CBE(M–X)/kJ mol^{-1}			
		F^-	Cl^-	Br^-	I^-
M(III)					
Sc	$3d^0$	1679	1531	1486	1444
Ti	$3d^1$	1801	1650	1608	1573
V	$3d^2$	1853	1698	1666	1631
Cr	$3d^3$	1869	1746	1708	1669
Mn	$3d^4$	1914	1795	1756	1721
Fe	$3d^5$	1881	1753	1727	1698
Co	$3d^6$	1965	1846	1811	1779
Ni	$3d^7$	2029	1910	1875	1846
Y	$4d^0$	1553	1405	1367	1325
La	$5d^0$	1486	1328	1280	1232
Complex M(III)					
$[FeCl_4]^-$	$3d^5$	–	1414	–	–
$[CrCl_6]^{3-}$	$3d^3$	–	900	–	–
M(IV)					
Ti	$3d^0$	2451	2280	2241	2200
Zr	$4d^0$	2178	2004	1964	1915
M(V)					
V	$3d^0$	3292	–	–	–
Nb	$4d^0$	2835	2655	2615	2578

4.4.2 Ligand Field Effects

The mean coordinate bond dissociation energies for the species MX_2, MX_3, and $[M^{II}X_4]^{2-}$ are plotted in Fig. 4.12 for the first transition series. For all three types the variation shows a similar humped pattern which corresponds to the number of *d*-electrons in the transition metal ion. For MX_2 and MX_3, the shapes of the curves are qualitatively in agreement with the predictions [11] of the crystal field theory on the assumption that MX_2 and MX_3 are linear and planar molecules, respectively.

For the complex tetrahalides a more detailed analysis is possible since we know (Table 4.3) the spectroscopically determined values of Δ_t for these 'high-spin' tetrahedral species. By subtracting stabilization energies consisting of one quarter (for each bond) of 0·6 Δ_t for Fe(II), 1·2 Δ_t for Co(II), and 0·8 Δ_t for Ni(II) from the mean $\overline{CBE}$ values we arrive at the 'corrected' dotted line in reasonable agreement with the CFT. The Cu(II) complex is

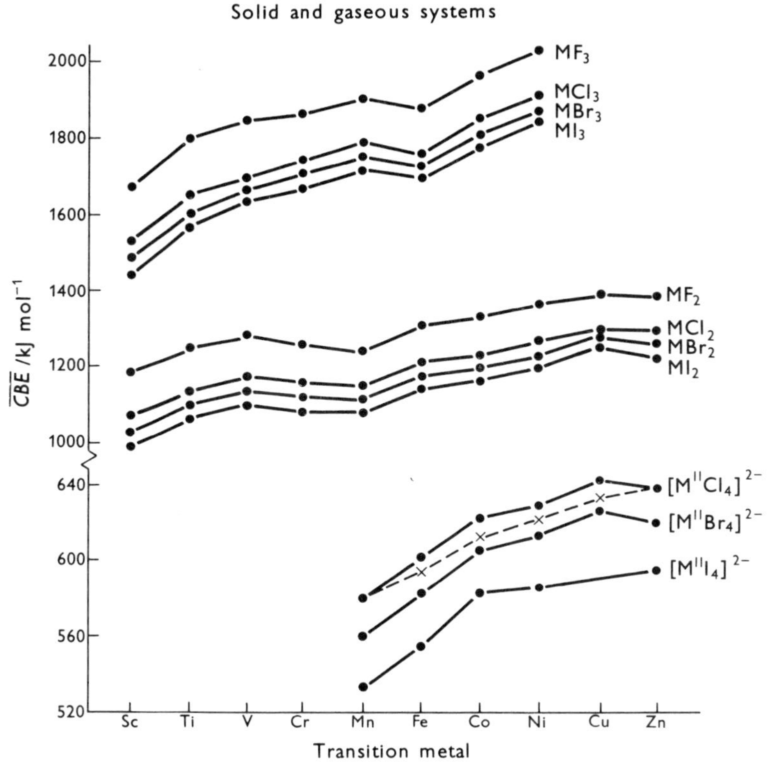

FIG. 4.12. Mean metal-halogen coordinate bond dissociation energies (from Table 4.6) for trihalides, dihalides, and complex tetrahalides of transition metals. Values shown (. . . × . . .) are 'corrected' for crystal field stabilization.

rather more stable than expected as we found for aqueous complexes (Section 4.3.1) and which we attributed to Jahn–Teller distortion.

Some thermodynamic aspects of tetrahalide complexes in non-aqueous solvents are considered in Section 7.2.3.

Strong support for the crystal field and molecular orbital analyses of metal-ligand interactions comes from the agreement between spectroscopic and thermochemical measurements of Δ_o in the cationic octahedral complexes $[M^{II}(H_2O)_6]^{2+}$ and $[M^{II}(NH_3)_6]^{2+}$ of the first transition series. The thermodynamic cycle needed for the calculation of the mean $M-H_2O$ and $M-NH_3$ coordinate bond dissociation energies is identical with that discussed in Section 2.4. As an illustration, $\overline{CBE}(M-NH_3)$ is plotted in Fig. 4.13 together with the $\overline{CBE}$s 'corrected' for crystal field stabilization using values of Δ_o for the 'high-spin' complexes from Table 4.3. The 'corrected' values are seen to lie on a smooth curve passing through the points for calcium, manganese, and zinc for which no stabilization exists.

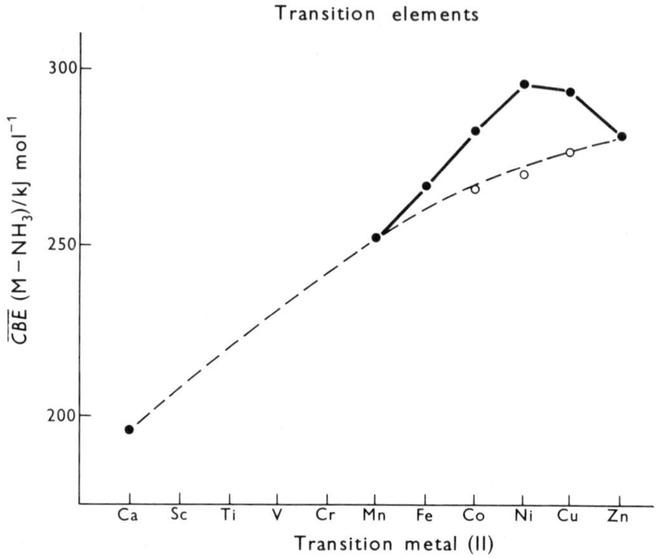

FIG. 4.13. Mean metal-ammonia coordinate bond dissociation energies [8] in $[M^{II}(NH_3)_6]^{2+}$ complexes in the gas phase. Values shown (○) are 'corrected' for crystal field stabilization.

It is interesting to compare the mean metal-ligand $\overline{CBE}$ values of 236 kJ mol^{-1} and 295 kJ mol^{-1}, respectively, for the hexaaquo and hexammine nickel (II) complexes with 363 kJ mol^{-1} for the sulphur-bonded octahedral complex hexathioureanickel (II) [13]. The bond energies increase in strength in the order of donor atom $0 < N < S$ showing the increasing softness of these ligands.

4.4.3 Theoretical Calculation of Bond Energies

The CFT has been seen to be quite successful in calculating the additional stabilization generated by the effect of the electrostatic ligand field on the metal *d*-orbitals, but it is in no way capable of allowing the calculation of the total *CBE*.

The *hard-sphere* model considers a molecule as an assembly of spherical, non-polarizable ions for which the bonding energy can be calculated from electrostatic considerations alone. It would be expected to be appropriate to ionic bonding, particularly in the gas phase where we have shown (Section 4.4.1) that hard behaviour is exhibited by all the transition metal ions. The interactions between pairs of ions, *i* and *j*, are given by

$$CBE = \sum_{i \neq j} z_i z_j \, e^2 / r_{ij}$$

where z_i and z_j are the charges on the ions, r_{ij} is the separation and e is the electronic charge. Actually terms are added to this formula to take into

Table 4.7. Mean Metal-halide Coordinate Bond Dissociation Energies Calculated by Various Theoretical Models [11].

Molecule	$\overline{CBE}$(M–X)/kJ mol^{-1} at 298 K			
	Experimental	Hard-sphere	Polarizable-sphere	MO
$CaCl_2$	970	878	950	979
$SrCl_2$	917	849	878	931
$BaCl_2$	864	806	825	897
$ZnCl_2$	1293	1013	1250	1250
$CdCl_2$	1182	917	1100	1140
$HgCl_2$	1283	902	1081	1225
$TiCl_4$	2280	1614	2120	2101
$[Zn^{II}Cl_4]^{2-}$	636	586	619	
$[Zn^{II}Br_4]^{2-}$	619	552	598	
$[Zn^{II}I_4]^{2-}$	594	502	548	

account longer range electrostatic interactions, and the overall calculation of *CBE* depends on the availability of experimental data on interionic distances and virial coefficients.

Values of *CBE* for several halides, calculated by the hard-sphere model are given in Table 4.7 in comparison with experimentally determined values. The compounds chosen for study are generally those for which no ΔH(CFSE) applies. For other compounds the ΔH(CFSE) could of course be calculated separately as we have done earlier in Section 4.4.2 if values of Δ are available. Agreement between calculated and experimental values of *CBE* is poor, but the differences are less marked for the harder, more ionic Ca^{2+}–Cl^- interaction than for the softer Hg^{2+}–Cl^- interaction in which covalent bonding might be expected to contribute significantly, even though these two metal ions both show hard behaviour in the gas phase. The difference between calculated and experimental values may indicate that covalent bonding is important even for the Ca^{2+}–Cl^- interaction in the gas phase. Actually, for bonding which would be expected to be almost entirely ionic, such as in the alkali metal halides which are outside the scope of this chapter, the agreement between theory and experiment is good, for example 557 kJ mol^{-1} and 555 kJ mol^{-1}, respectively, for NaCl.

An obvious improvement to the hard-sphere model is to consider the ionic spheres as being not hard but polarizable, the amount of polarizability depending on the electrostatic forces acting on the ion. The *polarizable-sphere* model adds terms to the hard-sphere *CBE* equations to allow for the production of induced dipoles on the halogen atoms by the central cation, terms for the interaction of the induced dipoles with other ions and with each other, and terms for van der Waals' type of interactions of ions of like and opposite charge.

The detailed calculations of these terms is outside the scope of this text but values calculated [8, 11] in this way are included in Table 4.7. Agreement between theoretical and experimental *CBE* values is now much better, emphasizing the importance of covalent interactions even in the gas phase.

A much improved *molecular orbital* model can be developed [11] by considering the construction for each M–X bond of a two-centre molecular orbital by the LCAO method (Section 1.3). A few values calculated by this method are included in Table 4.7 and are seen to be in good agreement with experiment.

4.5 Thermal and Chemical Stability of Di- and Trihalides

In earlier sections of this chapter, we have expressed the stabilities of both aqueous and solid compounds in terms of coordinate bond dissociation energies, since this is the form most readily related to bonding theories. When we wish to discuss thermal stability, chemical stability with respect to *elements*, the coordinate bond dissociation energy is inappropriate. We should rather discuss processes such as that shown below, for example for a dichloride

$$MCl_2(s) = M(s) + Cl_2(g) \qquad (4.3)$$

Alternatively, if we wish to gain some insight into the relative chemical stability of the solid di- and trihalides, a suitable theoretical reaction for study might be

$$MX_3(s) = MX_2(s) + 0{\cdot}5X_2(g) \qquad (4.4)$$

where X represents a halide ion. Values of $\Delta G^{\ominus}$ for this reaction across the first transition series, plotted in Fig. 4.14, are positive for scandium, titanium, vanadium and chromium, for all three halides Cl^-, Br^-, and I^-, indicating that the M(III) form is stable relative to the M(II) form. Indeed the values suggest that the dihalides, if formed, would disproportionate to MX_3 and metal. At manganese the value of $\Delta G^{\ominus}$ becomes negative and the trihalides become unstable relative to the dihalides. For iron, the trend is reversed for the chlorides and bromides, with the trihalide once more being the more stable, although uncertainties in the plotted values in the region Mn–Co are close to the values themselves. From cobalt onwards, the dihalides are the predominant species. These observations are generally in agreement with the chemical observations; for example, $ScCl_2$ and the trihalides from cobalt to zinc are unknown (fluorides are omitted from the calculations owing to a shortage of data).

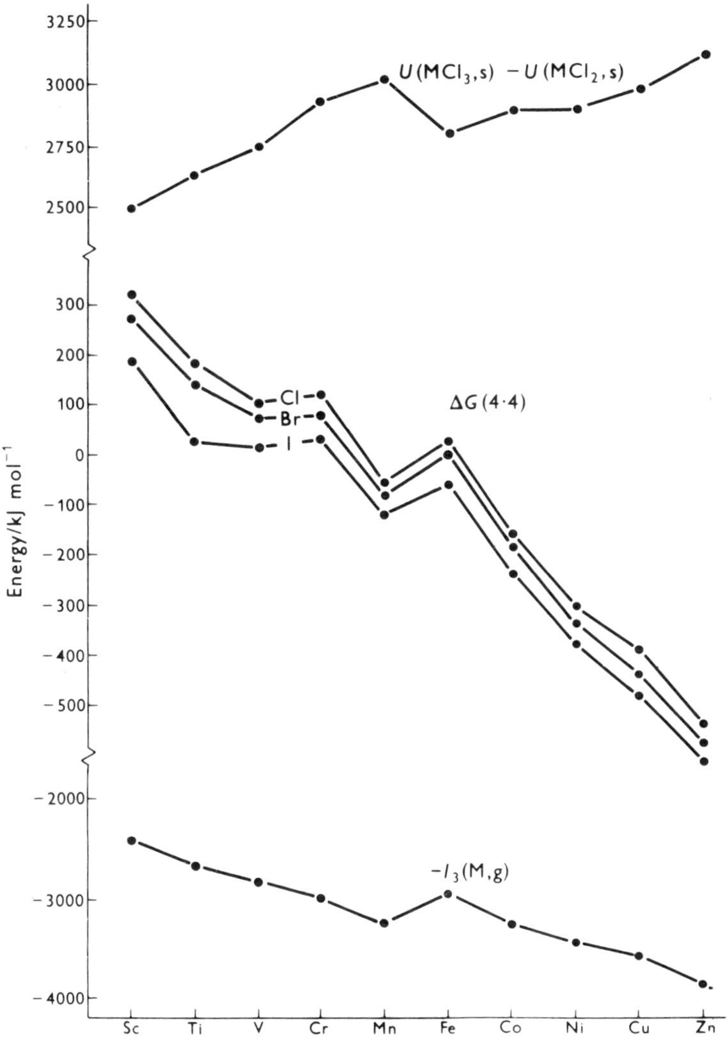

FIG. 4.14. Values [14] of $\Delta G(4.4)(\approx \Delta H)$ for halides. Also shown, for the chloride case, are the factors influencing the shape of the ΔG plot. The ionization energy term, $-I_3$, is much more important than the lattice energy term, U.

The calculation of the $\Delta G^\ominus$ values does not advance our understanding of why this order of stability is found; it merely restates the problem in terms of thermodynamically known quantities. A significant advance is possible, however, if $\Delta G^\ominus$ is split into $\Delta H^\ominus$ and $-T\Delta S^\ominus$ contributions, and the former are related by means of the cycle shown in Fig. 4.15 to a number of separate

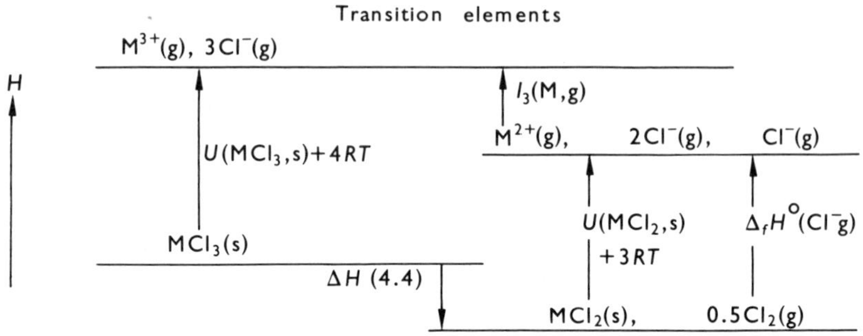

$$\Delta H(4.4) = -I_3(\mathrm{M,g}) + U(\mathrm{MCl_3,s}) - U(\mathrm{MCl_2,s}) - \Delta_f H^{\ominus}(\mathrm{Cl^-,g}) + RT$$

$$\Delta G(4.4) = \Delta H(4.4) - T\Delta S(4.4)$$

FIG. 4.15. Thermochemical cycle for relating together the values plotted in Fig. 4.14 for the chlorides. The term $\Delta_f H^{\ominus}(\mathrm{Cl^-, g})$ is constant for the series.

processes for which some theoretical advance has been made. Nelson and Sharpe [14] have carried out these calculations and the reader is referred to their paper for the data, some of which, of course, had to be estimated.

The main features of Nelson and Sharpe's calculations for all three halides Cl^-, Br^-, and I^-, are that the $-T\Delta S^{\ominus}$ terms have a negligible effect on the values of $\Delta G^{\ominus}$, and that the overriding influences on $\Delta H^{\ominus}$, and hence on $\Delta G^{\ominus}$, are the values of the third ionization energies, $I_3(\mathrm{M, g})$, and the two lattice energies $U[\mathrm{MX_3, s}]$ and $U[\mathrm{MX_2, s}]$. In the actual cycle the necessary corrections to convert these quantities to enthalpy changes should be made, as described in Section 2.2.2. It emerges that the *shape* of the ΔG° plot in Fig. 4.14 is determined by the variation in the third ionization energy. This can be seen from the plot of $-I_3$ values and lattice energies for the chloride case (the bromide and iodide are similar) in Fig. 4.14. An important feature of the $-I_3$ plot is the large discontinuity from manganese to iron, responsible for the chemical properties mentioned earlier. The lattice energies, although influencing by $U[\mathrm{MX_3}] - U[\mathrm{MX_2}]$ the overall value of $\Delta G^{\ominus}$, are relatively unimportant in determining the shape of the $\Delta G^{\ominus}$ plot.

It is the lattice energies, the energy changes from crystalline solid to gaseous ions, which include the crystal field stabilization energies, and Fig. 4.14 shows that these have the expected 'double-humped' shape (Section 4.4.2) with minima occurring at manganese for the M(II) species and at iron for the M(III) species. It is interesting that crystal field effects, so important in our earlier discussions of coordinate bond dissociation energies, make an insignificant contribution to the chemical and thermal stability of the first transition series halides.

Table 4.8. Electronic Configurations of Lanthanides and Actinides in their known Oxidation States. Radii of Trivalent Ions. (Data mainly from Reference [8])

Lanthanide		Atom			M(II)	M(III)	M(IV)	Radius M(III)/
	[Xe]+	4*f*	5*d*	6*s*	4*f*	4*f*	4*f*	nm
La			1	2		0		0·1061
Ce		1	1	2	2	1	0	0·1034
Pr		3		2		2	1	0·1013
Nd		4		2	4	3	2	0·0995
Pm		5		2		4		0·0979
Sm		6		2	6	5		0·0964
Eu		7		2	7	6		0·0950
Gd		7	1	2		7		0·0938
Tb		9		2		8	7	0·0923
	or	8	1	2				
Dy		10		2		9	8	0·0908
Ho		11		2		10		0·0894
Er		12		2		11		0·0881
Tm		13		2	13	12		0·0869
Yb		14		2	14	13		0·0858
Lu		14	1	2		14		0·0848

Actinide		Atom			M(III)	Radius M(III)/
	[Rn]+	5*f*	6*d*	7*s*	5*f*	nm
Ac			1	2		0·111
Th			2	2		
Pa		2	1	2		
	or	1	2	2		
U		3	1	2	3	0·103
Np		5		2	4	0·101
Pu		6		2	5	0·100
Am		7		2	6	0·089
Cm		7	1	2	7	0·098
Bk		8	1	2		
	or	9		2		
Cf		10		2		
Es		11		2		
Fm		12		2		
Md		13		2		
No		14		2		
Lw		14	1	2		

4.6 Lanthanides and Actinides

4.6.1 General

These two series are characterized by unfilled inner *f*-electron shells. Table 4.8 summarizes the electronic structures of both the neutral atoms and the M(III) ions, and also lists the radii of the M(III) ions. There is a decrease in the ionic radius along both the lanthanide series (the lanthanide contraction) and the actinide series which will be shown later to have important consequences in the thermodynamics of aqueous complexes. As in previous sections of this text, discussion of the thermodynamics of lanthanide and actinide compounds is limited to those aspects where adequate data are available; the antinides being very sparse in this respect. Much of this chapter will concern the coordination chemistry of the trivalent lanthanides in solution (Section 4.7) with a brief discussion (Section 4.8) of the thermal stability of the solid halides. Johnson [15] has dealt with the relative stability of the oxidation states of the lanthanides and actinides, but a full analysis is hindered by a lack of accurate thermodynamic data.

4.6.2 Bonding in Lanthanide Complexes

Whereas in the *d*-type transition series the *d*-electrons are able to take an important part in metal-ligand bonding, the lanthanide 4*f*-electrons are well shielded by electrons in 5*s* and 5*p*-orbitals from interactions with ligands, with the result that lanthanide complexes are of considerably lower stability. Stable complexes in the lanthanide series are formed generally with those ligands with the more electronegative donor atoms such as H_2O and edta.

Covalent bonding can only occur by participation of unoccupied higher-energy orbitals (for example, 5*d*, 6*s*, 6*p*) on the lanthanide ion, and might be expected only for ligands of strong coordinating ability such as the cyanide ion. In fact, oxygen-17 nuclear magnetic resonance measurements [16] on aqueous lanthanide ions were interpreted on the basis that only the 6*s*-orbitals were involved in covalent bonding in the hydrated ions, the 4*f*, 5*d*, and 6*p*-orbitals taking at most only a minor part in the bonding.

Electrostatic lanthanide-ligand interactions would be expected to be an important part of the bond energy. The bonding more closely resembles that in complexes of the non-transition metals such as calcium, strontium, and barium, rather than *d*-transition metal complexes. For this reason, the decreasing radius of the lanthanide (III) ions along the series should be a major influence on the variation in complex stability. We shall see later that other, less predictable factors modify this simple view.

4.6.3 Ligand Field Effects

The well-shielded position of the lanthanide 4*f*-orbitals and the minor role of these in bonding suggest that splitting of their energies by an electrostatic ligand field will be slight. Spectroscopic measurements of Δ_o for an octahedral field ($\approx$1·2 kJ mol^{-1}) indicate that ligand field effects may just be detectable in enthalpy data for complex formation. That this is possible, when ΔH is separated from a number of complicating factors, is shown in Section 4.7.2.

An octahedral field splits [17] the energies of the 4*f*-orbitals according to Fig. 4.16. The additional stabilization for 'high-spin' complexes is related to

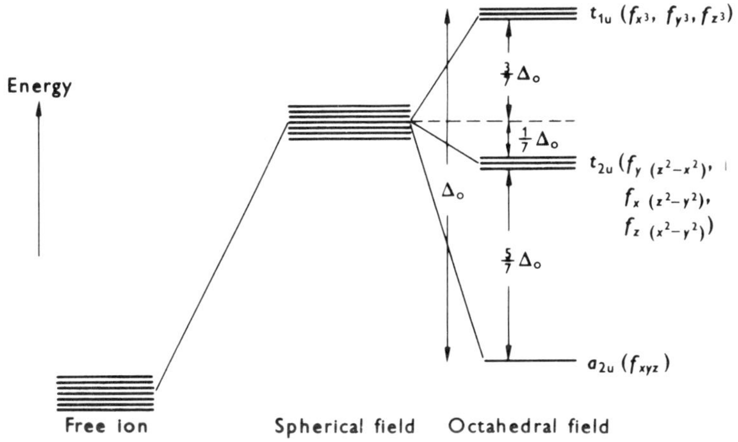

FIG. 4.16. Splitting of lanthanide 4*f* orbital energies in an octahedral crystal field.

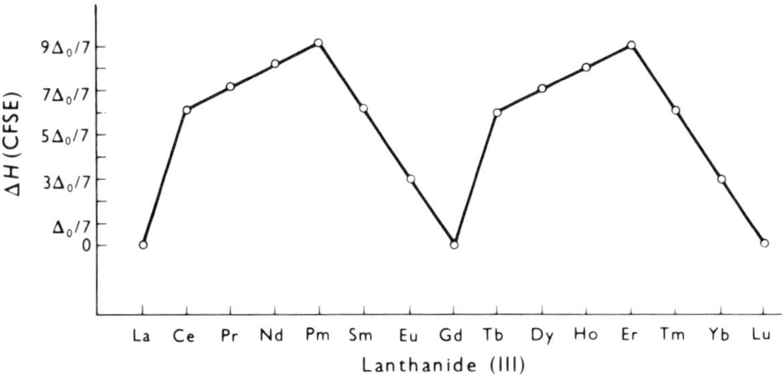

FIG. 4.17. Positive contributions of crystal field stabilization in an octahedral field to lanthanide-ligand *CBE* values.

the numbers, n, of electrons in the three groups of orbitals, a_{2u}, t_{2u}, and t_{1u}, by

$$\Delta H(\text{CFSE}) = \left[\frac{6}{7}n(a_{2u}) + \frac{1}{7}n(t_{2u}) - \frac{3}{7}n(t_{1u})\right]\Delta_o$$

resulting in the 'double-humped' stabilization curve shown in Fig. 4.17.

4.7 Aqueous Lanthanide Systems

4.7.1 Stability Across the Lanthanide Series

The variation in Gibbs function, ΔG, enthalpy, ΔH, and entropy, $-T\Delta S$, for the replacement of ligands by water for a number of lanthanide complexes is shown in Fig. 4.18. We have chosen to express the data in

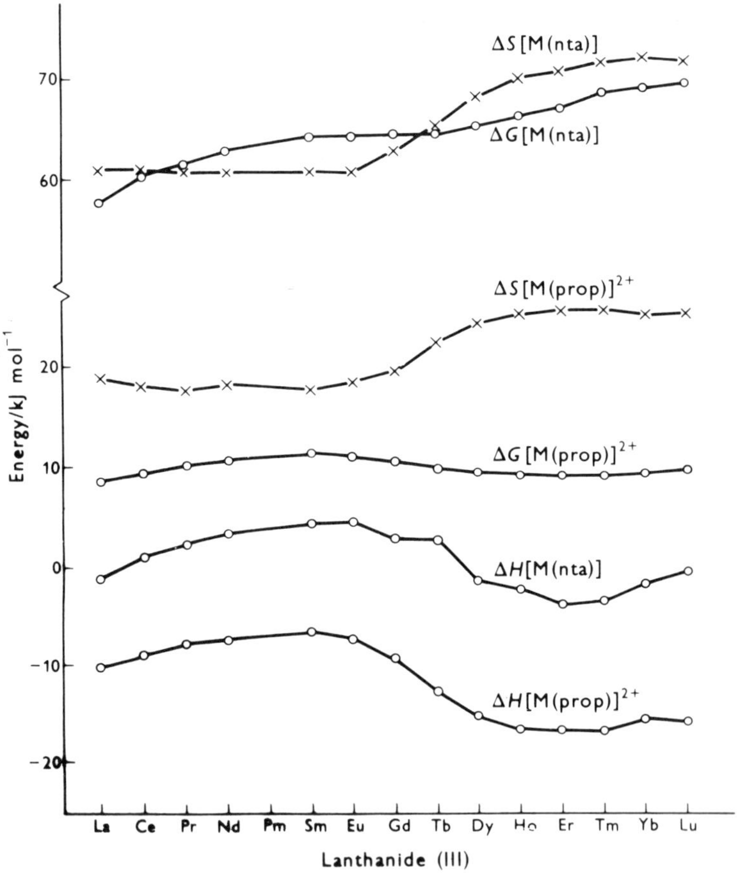

FIG. 4.18. Values [8] of ΔG, ΔH(= *CBE*), and $-T\Delta S$ for the replacement by water of the ligand in aqueous lanthanide propionate complexes (at 298 K, 2M $NaClO_4$) and nitrilotriacetate complexes (at 293 K, 0·5M $NaClO_4$).

terms of the metal-ligand bond-breaking reactions,

$$[MA_n]^{(3+nm-)} = M^{3+}(aq) + nA^{m-}(aq)$$

as we did in Section 4.3.1, so that the enthalpy values are analogous to the coordinate bond dissociation energies for gas-phase complexes. All the complexes represented in Fig. 4.18 are 1 : 1 lanthanide : ligand species with anionic ligands, bonding probably by predominantly electrostatic forces. It is interesting that the addition of further ligands does not greatly modify the shapes of the curves, as shown in Fig. 4.19 for the overall enthalpy changes for the bond-breaking of glycolate from successive lanthanide-glycolate complexes.

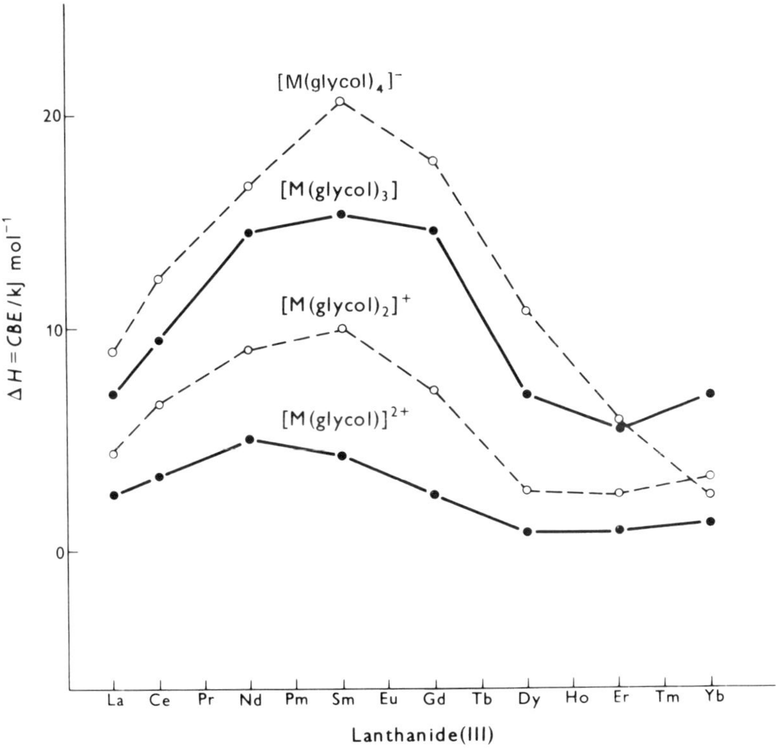

FIG. 4.19. Values [8] for the replacement by water of glycolate from aqueous lanthanide complexes $[M(glycol)_n]^{3+n-}$ at 298 K in 2M $NaClO_4$.

If purely electrostatic bonding were responsible for the stability of these complexes, then the enthalpy change (or *CBE*) should show a simple, almost linear variation along the series, determined, for a particular ligand, by the decreasing ionic radius of M(III) (Table 4.8).

The explanation [18] of the far-from-linear variation shown in Fig. 4.18 is now generally thought to be related to the changes in hydration sphere around the cations, as the series is traversed. Evidence for this comes from the ionic conductances which fall into two groups; those from lanthanum to gadolinium all have similar values but these differ markedly from the smaller values of the group from terbium to lutetium. This suggests different hydration spheres for the two groups of lanthanides, those of the heavier lanthanides having the larger hydration number. The effect of this should be twofold:

(1) the lighter lanthanides should require more energy to replace their ligands by water, relative to the heavier lanthanides, since fewer compensating M—H_2O bonds are formed in the former group with the smaller hydration sphere. This effect would be superimposed on the steadily increasing electrostatic metal-ligand interaction caused by the reducing lanthanide (III) radii along the series. The overall effect should be as shown in Fig. 4.20, which closely parallels the behaviour found experimentally.

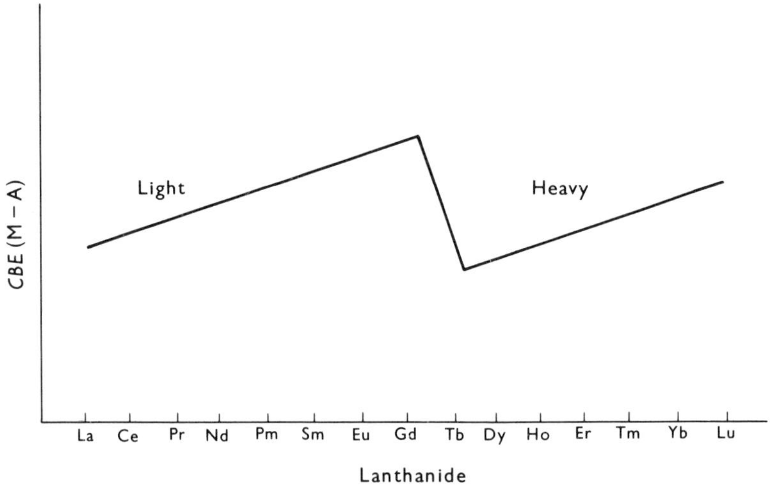

FIG. 4.20 Effect on lanthanide-ligand *CBE* values of an abrupt change of hydration number of lanthanide ions at Gd–Tb.

(2) the entropy values should be bunched together for the light lanthanides and for the heavy lanthanides. This is clearly seen in Fig. 4.18 in which, for the series of nitrilotriacetate (nta) complexes for example, the values of ΔS are -205 ± 1 J K^{-1} mol^{-1} from lanthanium to europium and -239 ± 8 J K^{-1} mol^{-1} from holmium to lutetium, with a transition occurring over the ions gadolinium, terbium, and dysprosium. The shape of the Gibbs function curve follows from $\Delta H - T\Delta S$.

4.7.2 Ligand Field Effects

The shape of the enthalpy curves in Figs 4.18 and 4.19 show no traces of the 'double-humped' pattern expected from our earlier discussion (Section 4.6.3) of the effect of an electrostatic field, although it was noted that the magnitude of such an effect was bound to be small. Also the relatively large values of ΔH caused by the changes in cation hydration around the centre of the series (see previous section) should effectively mask any more subtle ligand field effects.

To overcome the complication of the variable hydration of the lanthanide ions, Staveley, Markham, and Jones [19] have applied the following series of measurements to the metal-ligand bond-breaking reaction, illustrated here (reaction (4.5)) for the lanthanide-dtpa (an octadentate 5^- anion, see Table 4.4) complexes.

$$[M(dtpa)]^{2-}(aq) = M^{3+}(aq) + dtpa^{5-}(aq) \qquad (4.5)$$

These authors measured the enthalpies, $\Delta H(4.6)$ and $\Delta H(4.7)$, of solution of two series of crystalline lanthanide salts, the bromate enneahydrates and the ethylsulphate enneahydrates, each of which forms an isomorphous series, with all the water molecules coordinated to the lanthanide ion.

$$M(BrO_3)_3 9H_2O(s) = M^{3+}(aq) + 3BrO_3^-(aq) + 9H_2O(aq) \qquad (4.6)$$

$$M(EtSO_4)_3 9H_2O(s) = M^{3+}(aq) + 3EtSO_4^-(aq) + 9H_2O(aq) \qquad (4.7)$$

Subtraction of $\Delta H(4.6)$ or $\Delta H(4.7)$ from the enthalpy of reaction (4.5) eliminates the aqueous lanthanide ions from the overall conversion of the aqueous complex into crystalline lanthanide salt, for example the ethylsulphate in reaction (4.8).

$$[M(dtpa)]^{2-}(aq) + 3EtSO_4^-(aq) + 9H_2O(aq) = M(EtSO_4)_3 9H_2O(s) + dtpa^{5-}(aq) \qquad (4.8)$$

Of the quantities in this equation, the complex $[M(dtpa)]^{2-}$ and the crystalline lanthanide ethylsulphate should both be stabilized by the crystal field. Other quantities are constant for the complete series. Nothing remains which should cause an abrupt change in the value of $\Delta H(4.8)$.

Figure 4.21 shows a plot of $\Delta H(4.8)$ for aqueous lanthanide-dtpa complexes which is expected to be independent of the variable hydration of the lanthanide ions. The trend of increasing complex stability along the series may be related to the decreasing lanthanide ionic radius and hence increasing electrostatic attraction. Relative to the complexes of lanthanum, gadolinium and lutetium, for which no crystal field stabilization occurs, the remaining lanthanides show the expected 'double-humped' enthalpy variation with stabilization by 1 to 3 kJ mol^{-1} which may be attributed to the effect of the

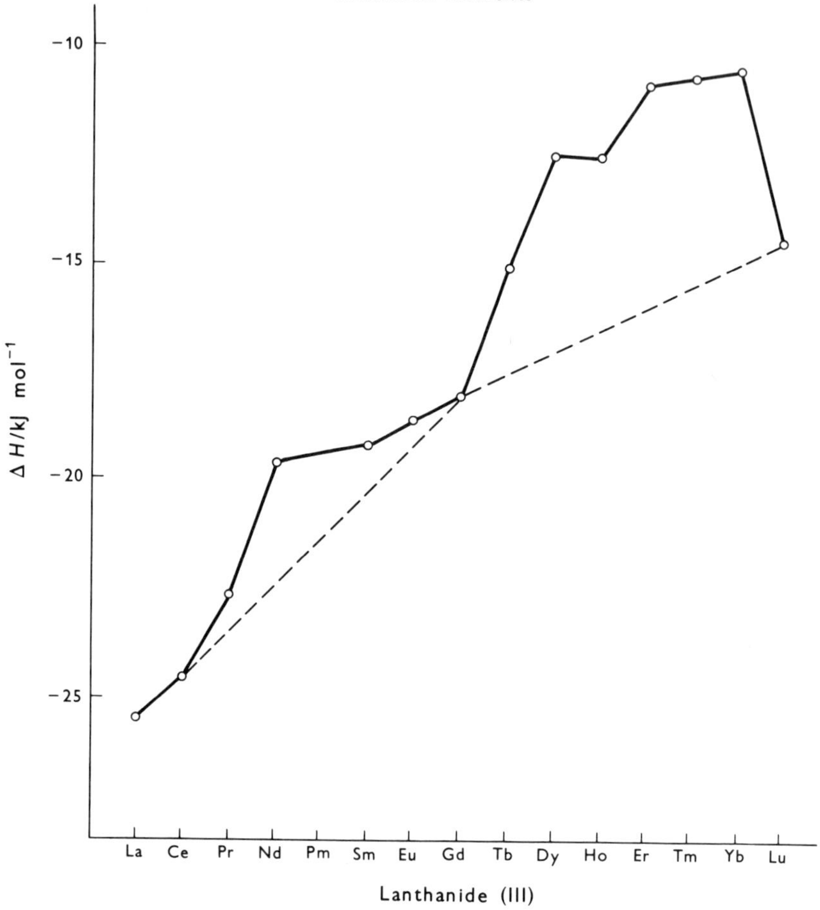

FIG. 4.21. Enthalpies [8, 19] of reaction (4.8) for the replacement by water of the ligand from aqueous lanthanide $[M(dtpa)]^{2-}$ complexes. Values are independent of the state of hydration of M, and they show crystal field effects.

crystal field. It is to be noted, however, that not all lanthanide complexes show this effect quite so clearly as does dtpa.

4.8 Solid Lanthanide Systems

4.8.1 Thermal and Chemical Stability of Chlorides

In our earlier discussion of the thermal stability of the di- and trihalides of the first transition series (Section 4.5), we mentioned that the appropriate bond energies were not the coordinate type, involving ionic species, but the dissociation into atoms. It is this type of process which is the more likely reaction in chemical and thermal situations at ordinary temperatures.

All the lanthanides exist as trihalides but a number are also stable as dihalides. Johnson [20] has calculated the enthalpy change for the disproportionation of the dichlorides

$$3MCl_2(s) = M(s) + 2MCl_3(s) \quad (4.9)$$

by estimating the thermodynamic properties of those dihalides which are unknown. It was noted when we discussed the thermal stability of the halides of the first transition series that the entropy change was negligibly small in comparison with the enthalpy change, and a similar assumption seems reasonable here. Thus $\Delta H(\approx \Delta G)$ for the above reaction should be a

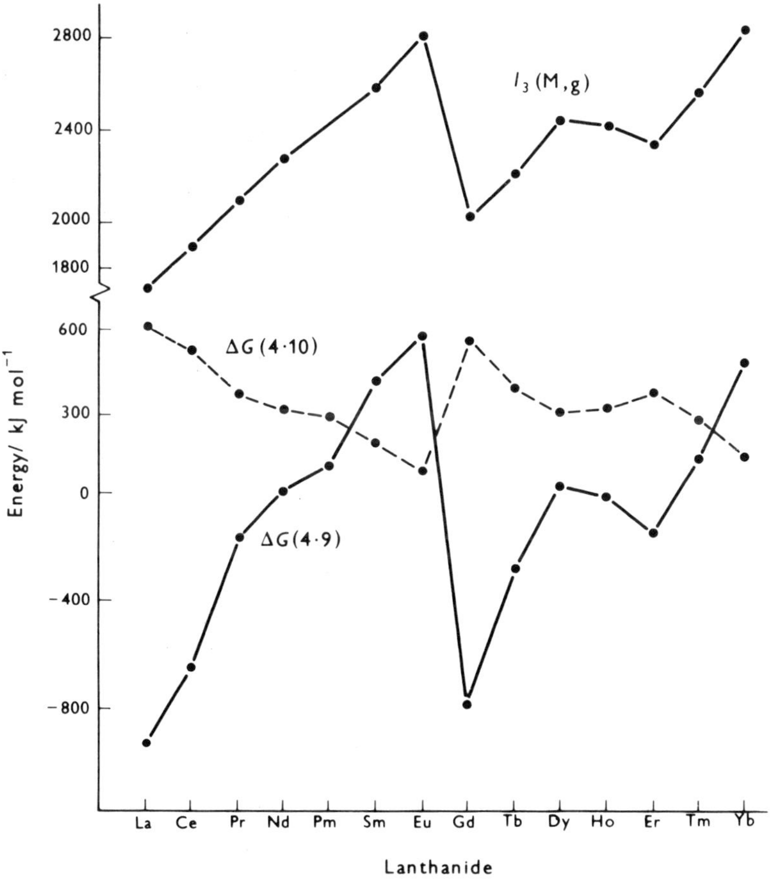

FIG. 4.22. Values [20] of $\Delta G(\approx \Delta H)$ for reactions (4.9) and (4.10). The major influence on these values is the lanthanide third ionization energy, I_3, which is also plotted.

guide to the stability of the lanthanide dihalides with respect to disproportionation at 298 K.

The values of $\Delta G(4.9)$, plotted in Fig. 4.22, indicate that the dihalides of lanthanum, cerium, praseodymium, gadolinium, erbium and terbium should disproportionate since ΔG is negative for these. This agrees well with chemical observations. Attempts to prepare these chlorides have so far failed, but the dichlorides of neodymium, dysprosium and thulium are known.

Calculations on the reaction (4.10) for the decomposition of the trichlorides to the dichlorides and gaseous chlorine,

$$MCl_3(s) = 0{\cdot}5Cl_2(g) + MCl_2(s) \qquad (4.10)$$

show that $\Delta G (\approx \Delta H)$ is positive in all cases. Decomposition of the trihalides is unlikely. This is also in agreement with the chemical observations.

The values of $\Delta G (\approx \Delta H)$ can be related to the properties of each metal by means of cycles similar to that in Fig. 4.15. It transpires, as it did for the first transition series, that the main contribution to the *trends* in ΔG for reactions (4.9) and (4.10) is the third ionization energy of the lanthanide metal, as shown in Fig. 4.22.

In these calculations, Johnson [20] emphasizes that the hypothetical dichlorides which he considered, involved M(II) ions of electronic configuration $5d^0\ 4f^{n-1}$, consistent with the known dihalides. It is possible that a different electronic configuration, such as $5d^1\ 4f^n$, could lead to a more stable dichloride than the type shown in Fig. 4.22.

REFERENCES

[1] Pauling, L., *The Nature of the Chemical Bond*, 3rd edn, Cornell Univ. Press, Ithaca (1960).
[2] Jaffé, H. H., and Orchin, M., *Symmetry in Chemistry*, Wiley, New York (1965).
[3] Guzzetta, F. H., and Hadley, W. B., *Inorg. Chem.* **3**, 259 (1964).
[4] Orgel, L. E., *An Introduction to Transition-Metal Chemistry*, Methuen, London (1960).
[5] Figgis, B. N., *Introduction to Ligand Fields*, Interscience, New York (1966).
[6] Cotton, F. A., and Wilkinson, G., *Advanced Inorganic Chemistry*, 3rd edn, Interscience, New York (1972).
[7] Gray, H. B., *J. Chem. Educ.* **41**, 2 (1964).
[8] Ashcroft, S. J., and Mortimer, C. T., *Thermochemistry of Transition Metal Complexes*, Academic Press, London (1970).
[9] National Bureau of Standards, *Technical Note* 270–4, Washington D.C. (1969).
[10] Buckley, R. R., and Mercer, E. E., *J. Phys. Chem.* **70**, 3103 (1966).
[11] Pearson, R. G., and Mawby, R. J., In *Halogen Chemistry* Vol. 3 (edited by V. Gutmann), Ch. 2, Academic Press, London (1967).
[12] Ahrland, S., *Structure and Bonding*, **5**, 118 (1968).
[13] Ashcroft, S. J., *J. Inorg. Nuclear Chem.* **33**, 603 (1971).

[14] Nelson, P. G., and Sharpe, A. G., *J. chem. Soc.* (*A*) 501 (1966).
[15] Johnson, D. A., *J. chem. Soc.* (*A*) 1528 (1969).
[16] Lewis, W. B., Jackson, J. A., Lemons, J. F., and Taube, H., *J. chem. Phys.*, **36**, 694 (1962).
[17] Yatsimirskii, K. B., and Kostromina, N. A., *Russ. J. Inorg. Chem.* **9**, 971 (1964).
[18] Choppin, G. R., and Graffeo, A. J., *Inorg. Chem.* **4**, 1254 (1965).
[19] Staveley, L. A. K., Markham, D. R., and Jones, M. R., *J. inorg. nucl. Chem.* **30**, 231 (1968).
[20] Johnson, D. A., *J. chem. Soc.* (*A*), 2578 (1969).

SUGGESTIONS FOR FURTHER READING

Bonding and Group Theory

Cotton, F. A., *Applications of Group Theory*, Wiley, New York (1963).
Dunn, T. M., McClure, D. S., and Pearson, R. G., *Some Aspects of Crystal Field Theory*, Harper and Row, New York (1965).
Moeller, T., *The Chemistry of the Lanthanides*, Reinhold, New York (1963).
Murmann, R. K., *Inorganic Complex Compounds*, Reinhold, New York (1964).
References [2, 4, 5, 6].

Thermodynamics of Complexes

George, P., and McClure, D. S., *Progr. inorg. Chem.* **1**, 381 (1959).
Rossotti, F. J. C., In *Modern Coordination Chemistry* (edited by J. Lewis and R. G. Wilkins), Ch. 1, Interscience, London (1960).
Nancollas, G. H., *Interactions in Electrolyte Solutions*, Elsevier, Amsterdam (1966).
Beech, G., *Q. Rev. chem. Soc.* **23**, 410 (1969).
References [8, 12].

Thermodynamic Data on Complexes

Sillén, L. G., and Martell, A. E., *Stability Constants of Metal-Ion Complexes*, Chem. Soc. Spec. Pub. No. 17 (1964), and supplements.
Christensen, J. J., and Izatt, R. M., *Handbook of Metal-Ligand Heats*, Dekker, New York (1970).
Reference [8].

Chapter 5

Metal–carbon Bonds

5.1 Introduction

Both the transition elements and the non-transition elements form a wide range of compounds containing metal- (or metalloid-) carbon bonds. In the type usually known as organometallic, the metal is bonded to the carbon atom of an organic group such as an alkyl or aryl radical. Metal-carbon bonds are also present in the carbonyls, in which the bonding is to the carbon monoxide molecule. We shall consider both types of compound in this chapter, since there is no clear difference between the bonding in the two groups, but we shall concentrate on those limited areas of the subject where thermochemical data are available.

In general, the non-transition elements (including the metalloids such as boron, silicon, phosphorus, arsenic, etc.) all form simple alkyl and aryl compounds, MR_n, in which the bonding, except for alkali-metal alkyls such as Na^+R^-, is predominantly of the σ-covalent type (Section 1.3). The appropriate thermodynamic quantity for discussing the factors affecting the stability of these compounds is the bond dissociation energy, namely the value of ΔH for the dissociation of the gaseous compound into gaseous uncharged metal and alkyl radicals, all in their ground states.

$$MR_n(g) = M(g) + nR(g)$$

$\Delta H/n = \bar{D}(M\text{–}R)$, the mean metal–alkyl bond dissociation energy. Usually, values of ΔG and ΔS for the same process are not known but are relatively unimportant in comparison with ΔH (see Section 2.3.3).

The fragments produced by dissociation of a particular bond in a molecule are frequently different in shape and electronic structure from their form in the undissociated molecule. The name *intrinsic bond dissociation energy* is given to the enthalpy change for a hypothetical dissociation process in which the fragments retain the shape and electronic structure which they had in the undissociated molecule. Unfortunately, the information needed to calculate the energy of the excited valence state of metals and molecules relative to their ground state is generally not available, or is of a very approximate nature. Therefore, although it may be preferable to consider intrinsic bond dissociation energies when comparing the strengths of bonds, mean bond dissociation energies $\bar{D}$(M–R) are often the only available data. We shall make further mention of the intrinsic quantities in a number of places throughout this chapter.

In contrast to the non-transition elements, the transition elements only rarely form simple alkyl and aryl compounds, except in the presence of other 'supporting' ligands such as tertiary phosphines. Bond dissociation energies in these species are less readily obtained since it is not always easy to apportion enthalpy changes between the several types of bond which may be broken in a reaction. In addition, fewer measurements have been made on the transition series organometallic compounds.

Transition elements also form a number of compounds with unsaturated organic molecules such as alkenes, cyclopentadienyl, and benzene, and with carbon monoxide in which the bonding may be a combination of σ and π types. A useful semi-empirical rule to determine the formula (and sometimes the structure) of transition metal organometallic compounds is that the total number of valence electrons is usually eighteen, arising from the number of valence electrons on the uncomplexed metal atom or ion plus the number of electrons contributed to the bonding by the ligands. An adjustment is made for the charge, if any, on the final molecule. Table 5.1

Table 5.1. Classification of Carbon-containing Ligands on the Basis of the Numbers of Electrons Contributed to M–C Bonding

Number of electrons	Examples of ligands
1	alkyl, aryl (R·)
2	alkenes, carbon monoxide, alkynes
3	π-allyl (CH_2=CH–CH_2·)
4	butadiene (CH_2=CH–CH=CH_2)
5	π-cyclopentadienyl (C_5H_5·)
6	π-benzene (C_6H_6)

lists some carbon-containing ligands in order of increasing numbers of electrons available on the ligand for bonding. This forms a useful classification of organometallic compounds which we shall follow in the remainder of this chapter.

The reader is referred to the short book by Rochow [1] and the comprehensive survey by Coates, Green, and Wade [2] for further details of organometallic chemistry.

Table 5.2. Mean Bond Dissociation Energies, $\bar{D}$(M–C)/kJ mol^{-1} at 298 K, for Alkyl and Aryl Compounds. (Data are from References [3] and [4], and from Table 5.3.)

Compound	Me	Et	*n*Pr	*i*Pr	*n*Bu	*i*Bu	*s*Bu	Ph	Bz
Group IA									
LiR		210			257				
Group IIB									
ZnR_2	180	144	159	162					
CdR_2	144	110							
HgR_2	126	101	102	84	118	109		136	
Group IIIB									
BR_3	371	346	354	345	354	340	320	446	
AlR_3	280	271							
GaR_3	250								
InR_3	167								
Group IVB									
CR_4	359	343							
SiR_4								313	
GeR_4	270	243	239					289	188
SnR_4	223	194	200	180	202			234	
PbR_4	160	129							
Group VB									
PR_3	279	229						297	
AsR_3	239	190						266	
SbR_3	219	179						244	
BiR_3	147	105						177	
Group VIB									
SeR_2		250							

Abbreviations for radicals

Me = $CH_3\cdot$	methyl
Et = $C_2H_5\cdot$	ethyl
*n*Pr = $C_3H_7\cdot$	normal propyl
*i*Pr = $(CH_3)_2CH\cdot$	isopropyl
*n*Bu = $CH_3CH_2CH_2CH_2\cdot$	normal butyl
*i*Bu = $(CH_3)_2CHCH_2\cdot$	isobutyl
*s*Bu = $CH_3CH_2\dot{C}HCH_3$	secondary butyl
*t*Bu = $(CH_3)_3C\cdot$	tertiary butyl
Ph = C_6H_5	phenyl
Bz = $C_6H_5CH_2\cdot$	benzyl

5.2 Alkyls and Aryls: One-Electron Ligands

5.2.1 Bonding

Bonding of alkyl and aryl groups to metals involves no new principles beyond those described in Chapter 1. In the majority of cases, the bonding is predominantly σ-covalent using appropriate carbon sp^3 or sp^2 hybrid orbitals and metal σ-orbitals, apart from the alkali metal organometallics, some of which, such as *n*-butylsodium, are usually regarded as being salts, $Na^+C_4H_9^-$. In some of the transition metal aryls there are additional possibilities for π-bonding (Section 5.2.3).

5.2.2 Thermochemistry of Alkyls and Aryls of Non-transition Elements

Amongst the wide range of non-transition metal organometallic compounds available, several types, such as the 'silicone' polymers

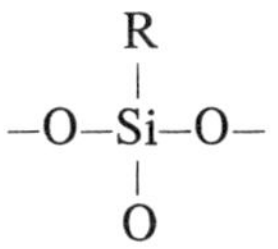

and the 'antiknock' tetraethyllead additives to fuel, have found commercial application. Many of the alkali metal and alkaline earth metal compounds (such as organolithium and Grignard reagents) are useful in organic syntheses. Only for a small fraction of these compounds are thermodynamic data available. We have expressed these data in Table 5.2 as mean metal-carbon bond dissociation energies, arranged in group order of metal (see Periodic Table (4.1)). These bond energies are related to the enthalpies of formation of the crystalline (or in some cases liquid) organometallic compounds by the

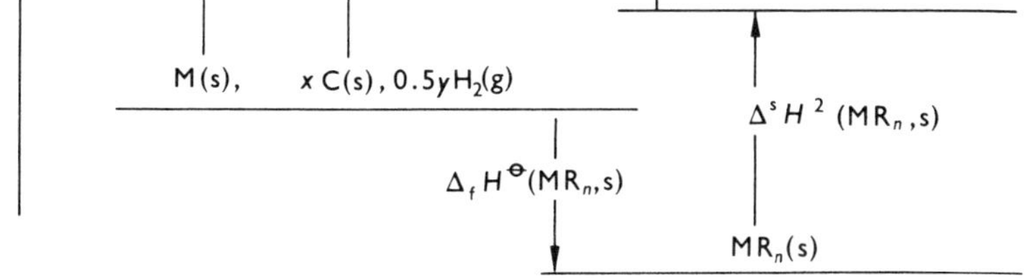

FIG. 5.1. Thermochemical cycle for calculating mean metal–carbon bond dissociation energies in metal alkyls and aryls, $MR_n (R = C_xH_y)$.

cycle shown in Fig. 5.1. The values of $\Delta_f H^{\ominus}(MR_n, s)$ and for $\Delta_s^g H^{\ominus}(MR_n, s)$, taken mainly from the tables of Cox and Pilcher [3], are most accurately found by rotating-bomb calorimetry (see Section 2.5.3). Values for silicon alkyls and aryls show considerable variation, possibly owing to the use of static bomb calorimetry for these compounds. Until more accurate data become available, we have used an approximate value for $SiPh_4$ in Table 5.1. Variation is so great for $SiMe_4$ and $SiEt_4$ that we have not included bond energies for these compounds.

Table 5.3. Standard Enthalpies of Formation of some Gaseous Atoms and Radicals. (Data are mainly from References [5] and [6]; see Table 5.2 for formulae of radicals.)

Gaseous atom	$\Delta_f H^{\ominus}$/ kJ mol^{-1}	Gaseous atom	$\Delta_f H^{\ominus}$/ kJ mol^{-1}	Gaseous radical	$\Delta_f H^{\ominus}$/ kJ mol^{-1}
H	218	Bi	207	Me •	142
Li	161	Se	227	Et ◦	108
Na	108	Sc	382	*n*–Pr •	88
Mg	147	Ti	471	*i*–Pr ◦	74
Zn	131	V	515	*n*–Bu •	71
Cd	112	Cr	397	*i*–Bu •	59
Hg	61	Mn	281	*t*–Bu •	28
B	563	Fe	416	*s*–Bu •	52
Al	326	Co	425	Ph •	301
Ga	277	Ni	430	Bz •	188
In	243	Cu	338	$C_3H_5^{\bullet}$ [a]	142
C	717	Mo	658	$C_5H_5^{\bullet}$ [b]	190
Si	456	Pd	378	$SnPh_3^{\bullet}$	532
Ge	376	W	849	$SnMe_3^{\bullet}$	95
Sn	302	Pt	565		
Pb	195	F	79		
P	314	Cl	122		
As	303	Br	112		
Sb	262	I	107		

a = allyl: *b* = cyclopentadienyl.

The values used for the standard enthalpies of formation of gaseous atoms and radicals are listed in Table 5.3, since some variation exists for these data in the literature.

Relative stability down each group

A noticeable trend is a decrease in $\overline{D}$(M–C) down each group, indicating that any reaction which involves the breaking of the metal-carbon bonds into atoms or radicals (rather than ions) should become increasingly easy down the group. Evidence from chemical reactivity and thermal stability confirms this. For example, chlorination of the tetraalkyls of Group IVB becomes increasingly easy from CEt_4 to $PbEt_4$, the higher members producing monochlorinated compounds, retaining metal-carbon bonds, while the lower members are completely broken down to chlorinated compounds without metal-carbon bonds.

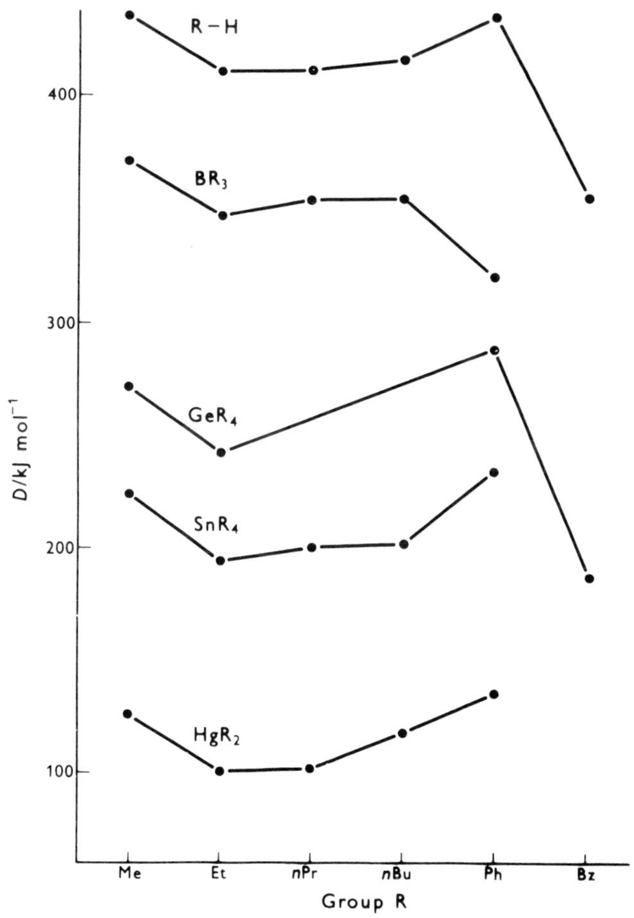

FIG. 5.2. Mean metal–carbon bond dissociation energies in organometallic compounds compared with D(R–H) in hydrocarbons.

The thermal stability of Group IVB tetraalkyls also decreases in the same order as the reducing M–C bond energies. Lead tetraalkyls most easily produce free alkyl radicals, a characteristic which is utilized in the 'anti-knock' properties of PbR_4. The free alkyl radicals are particularly effective as terminators of chain reactions.

Effect of different alkyl or aryl groups

The values of the mean bond dissociation energies decrease, on the whole, with increasing number of carbon atoms in the group R. Figure 5.2 shows that this trend parallels the bond dissociation energies for the breaking of carbon-hydrogen bonds in the equivalent alkane, suggesting similarities between the bonding in the two cases. Bonding of a metal to a phenyl group in some cases is rather stronger than expected from the value of D(Ph–H). There are possibilities of additional metal-phenyl π-bonding using appropriate metal p- or d-orbitals and the π-orbitals of the phenyl group which, since such bonding must be absent from the Ph–H bond, may account for the increased stability of these metal-phenyls.

The exceptionally weak germanium-benzyl bond, relative to the other members of the series, is a consequence of the particular stability of the benzyl radical which results from the extended π-orbital system available to the unpaired electron. The value of D(Bz–H) also shows this effect.

Stepwise bond dissociation energies

It should be noted that the thermochemical method outlined above gives only the total or mean bond dissociation energies. The stepwise removal of the alkyl groups in general involves enthalpy changes which differ considerably from the mean values owing to major rearrangements of the electronic structures of the progressively de-alkylated species in comparison with the electronic structure of the original molecule. The only requirement from a thermochemical point of view is that the sum of the individual or stepwise bond dissociation energies should equal the total bond dissociation energy determined as previously described.

Stepwise bond dissociation energies, of which a number are shown in Table 5.4, are available from studies of the kinetics of decomposition of the metal-alkyls, either thermally or by electron bombardment in a mass spectrometer. It is frequently reasonable to assume that the *recombination* of radicals involves little or no activation energy so that the activation energy, measured kinetically, for the *decomposition* reaction will be equal to ΔH for the same process (when appropriate RT terms have been added, see Section 2.2.2).

The most striking feature of the bond dissociation energies in Table 5.4 is the large difference between D_1 and D_2 for the dialkyls. To see how some of

Table 5.4. Stepwise and Mean Bond Dissociation Energies in Alkyls and Aryls [4, 7]

Compound	D_1/kJ mol^{-1}	D_2/kJ mol^{-1}	$\bar{D}$/kJ mol^{-1}
$ZnMe_2$	197	163	180
$CdMe_2$	192	96	144
$HgMe_2$	215	37	126
$HgEt_2$	178	24	101
$Hg(i\text{–}Pr)_2$	113	55	84
$SnMe_4$	257		223
$SnPh_4$	262		234
$SbMe_3$	238		219
$BiMe_3$	184		147

this difference arises, recall that in the introduction to this chapter we briefly mentioned intrinsic bond dissociation energies. The relation between intrinsic bond dissociation energies, D^*, the normal bond dissociation energies, D, and the value of Δ^*H necessary for the production of M and R in their valence states from their ground states, is shown in the cycle of Fig. 5.3. The situation

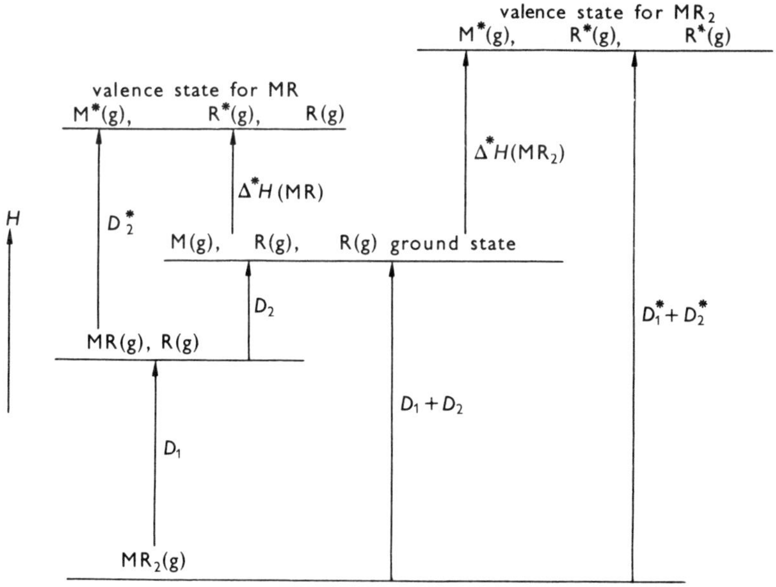

FIG. 5.3. Thermochemical cycle for the dissociation of an alkyl- or aryl-metal compound, MR_2.

is considerably complicated by the fact that the values of Δ^*H will not be the same for the species MR_2 and MR, nor will the bonding be necessarily of the same type in these two species. The relationship connecting D_1 and D_2 is complicated and has not been solved for metal alkyls and aryls.

For the alkyls in general, $D_1 > \overline{D}$.

Metal–metal bonds in Group IVB

A particular feature of the organometallic compounds of this group is the occurrence of metal–metal bonds as in $R_3M{-}MR_3$ for example. At present, data are scarce for the strength of these bonds but from the enthalpies of formation of the radicals given in Table 5.3 together with the enthalpies of formation of the binuclear compounds [4], $R_3M{-}MR_3$, we calculate that $D(Sn{-}Sn) = 223$ kJ mol^{-1} in $Me_3Sn{-}SnMe_3$ and 215 kJ mol^{-1} in $Ph_3Sn{-}SnPh_3$. Similarly, $D(C{-}C)$ in $Me_3C{-}CMe_3$ is 270 kJ mol^{-1}; the value of $\Delta_f H^\ominus(C_2Me_6, g)$ must be estimated, but this is quite accurately done for organic compounds [3]. On the assumption that $\overline{D}(Ge{-}Et)$ is unchanged from $GeEt_4$ to Ge_2Et_6, we calculate that $D(Ge{-}Ge) = 256$ kJ mol^{-1}.

The stability of metal–metal bonding increases from a rather weak Sn–Sn bond to the strong C–C bond. This parallels the observed chemical behaviour in which Sn–Sn and Pb–Pb bonds are readily broken thermally or by chemical reaction. The stable C–C bonds form the basis of a large number of organic compounds. Chemically, germanium shows an intermediate behaviour between carbon and tin with respect to the stability of metal–metal bonding.

5.2.3 Transition Elements

The transition elements form relatively few simple alkyls and aryls. Those which are formed tend to be less stable to decomposition to metal and hydrocarbon than corresponding non-transition element organometallics, although there seems to be no reason to expect any fundamental differences in bonding between the two series. For example, tetramethyltitanium is stable in diethylether solution only below 233 K whereas tetramethyltin is stable at room temperature.

A much wider range of alkyl- and aryl-transition metal complexes, particularly of the platinum group, is formed when other ligands are also present. At one time, trialkyl and triarylphosphine molecules were the preponderant supporting ligands, leading to the theory that the metal-carbon σ-bond was stabilized by the π-acceptor properties of the phosphines. This was supported by a molecular orbital calculation which showed poor overlap between ligand and metal d-orbitals. However, the variety of supporting ligands now recognized casts doubt on the assumption of inherently weak transition metal-carbon bonding, and evidence mentioned in this section certainly points to strong transition metal-carbon bonds. The presence of

$d\pi$-orbitals on the metal would be expected to add to the strength of the metal-carbon bond when the ligand has suitable π-orbitals for interaction, as is the case for aryl groups.

The only thermodynamic measurement [8] relates to the following reaction

$$trans\text{–}Pt(PEt_3)_2Ph_2(s) + HCl(dioxan) = trans\text{–}Pt(PEt_3)_2PhCl(dioxan) + C_6H_6(dioxan)$$

for which ΔH was found calorimetrically to be -128 kJ mol^{-1}. Subsidiary measurements enabled the enthalpy of the same reaction, but entirely in the gas-phase, to be calculated as -120 kJ mol^{-1}. To proceed, it is necessary to assume that D(Pt–Cl) is close to $0{\cdot}5\Delta H$ for

$$PtCl_2(g) = Pt(g) + 2Cl(g) \quad \Delta H \approx 779 \text{ kJ mol}^{-1}$$

Data from Reference [5] and Table 5.3 give D(Ph–H) = 435 kJ mol^{-1} and D(H–Cl) = 432 kJ mol^{-1}, leading to D(Pt–Ph) = 272 kJ mol^{-1}; This is far from a weak bond, being of the magnitude for example of the Ge–Ph bond in $GePh_4$, $\bar{D}$ = 289 kJ mol^{-1} (Table 5.2).

In the absence of thermodynamic data, an indication of bond strength may be obtained from spectroscopic infrared stretching absorption frequencies for M–C bonds, increasing frequency indicating increasingly strong bonding. For a variety of platinum complexes, these lie in a narrow range, implying that many organometallics in addition to the one studied thermochemically also have strong bonds. Similarly, there is very little variation in Pt–C bond lengths of $\approx 0{\cdot}210$ nm despite the diversity of supporting ligands.

It is interesting that, although as noted earlier $TiMe_4$ is less thermally stable than $SnMe_4$, there is very little difference in the force constants as determined from the infra-red stretching frequency for the M–C bonds. $SnMe_4$ has a mean M–C bond dissociation energy of 223 kJ mol^{-1} (Table 5.2), implying a similar value for $TiMe_4$.

It seems likely that the greater reactivity of some transition metal organometallics may be partly a problem of kinetic rather than thermodynamic stability, although it must be emphasized that it is not always clear that comparisons are being made on the basis of similar decomposition processes in the transition and non-transition cases. This illustrates the caution with which measurements of stability should be taken.

A detailed analysis of the possible pathways and kinetic factors for the breaking of metal-carbon bonds has recently been given by Braterman and Cross [9], to whose paper the reader is referred for details.

5.3 Carbon Monoxide Complexes: A Two-Electron Ligand

Most transition elements form carbonyls in which it is known, from X-ray structural analyses, that the carbon monoxide molecule is bonded through the carbon atom, and that the M—CO group is linear. Many carbonyls are important intermediates in industrial chemical processes and some show catalytic activity. Physically, they are generally flammable, sometimes toxic, liquids or solids which are soluble in non-polar solvents. From the point of view of the M—C bonding, they show similarities to the alkenes which are also two-electron ligands. The latter are discussed in Section 5.4.

The more common carbonyls are mononuclear, such as $Fe(CO)_5$, or binuclear, such as $Mn_2(CO)_{10}$ or $Fe_2(CO)_9$, but for metals of the platinum group, other ligands such as phosphines are usually also bonded to the metal.

The '18 valence-electron rule' (Section 5.1) is a simple rule for rationalizing the formulae of the carbonyls, but it is not infallible. In $Fe(CO)_5$ there are eight valence electrons plus two from each carbonyl, making a total of eighteen. For atoms initially having an odd number of electrons (for example, manganese and cobalt), any mononuclear carbonyl would also have an odd number of electrons. In these cases, a binuclear carbonyl, containing a metal—metal bond, is usually formed. Each metal is considered to donate one electron to the other. The structure of dimanganese decacarbonyl is thus $(CO)_5Mn—Mn(CO)_5$ in which each Mn is in an 18-electron environment. In contrast, the carbonyl $Fe_2(CO)_9$ requires three bridging CO molecules and an Fe—Fe bond to conform to the 18-electron rule. This is confirmed by X-ray structure analysis:

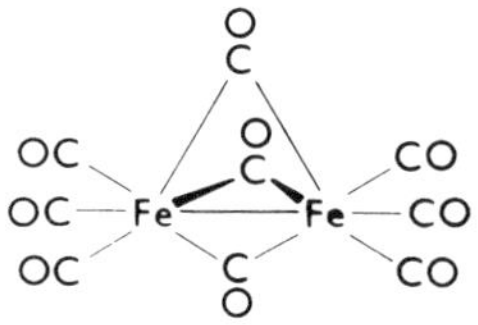

An exception to the 18-electron rule is the 17-electron $V(CO)_6$ where, possibly, steric factors prevent dimerization.

5.3.1 Bonding in Transition Metal Carbonyls

The electronic structure of carbon is $1s^2\ 2s^2\ 2p^2$ and that of oxygen $1s^2\ 2s^2\ 2p^4$. The bonding in carbon monoxide is considered to be a σ-bond between a hybrid carbon *sp*-orbital and an oxygen $2p_x$ orbital (Fig. 5.4a), leaving lone pairs of electrons on carbon (in an *sp*-orbital) and on oxygen (in a 2*s*-orbital). A doubly degenerate π-bond is also formed between the remaining carbon and oxygen $2p_y$ and $2p_z$ orbitals which make use of the four available electrons (Fig. 5.4b). Actually, bond lengths and dipole moments in CO and CO_2 indicate that the $2p_x$ oxygen orbital involved in the C—O σ-bond must have a little 2*s* character.

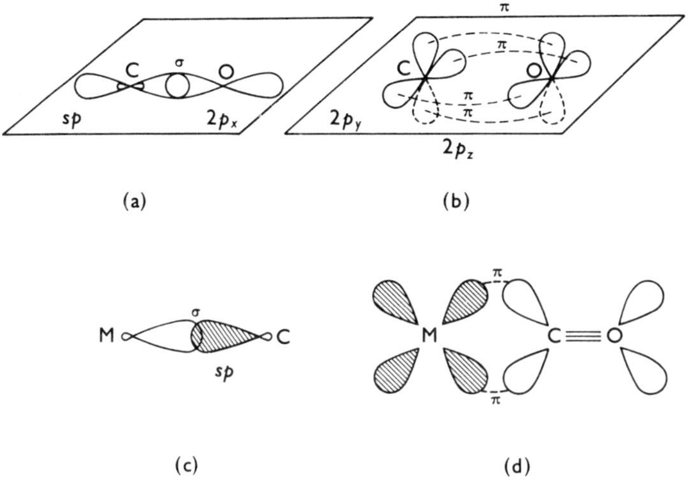

FIG. 5.4. Bonding in carbon monoxide: (a) and (b), and in metal carbonyls: (c) and (d).

The bonding between a transition metal and CO is firstly a dative overlap of the filled carbon *sp*-orbital with a suitable metal σ-orbital (Fig. 5.4c), and secondly a dative overlap of a filled metal *dπ*-orbital with an empty *pπ** antibonding orbital on CO (Fig. 5.4d). The back π-bonding helps to reduce a possible build-up of negative charge on M caused by the C→M interaction. One result of the formation of a strong M–CO bond is a reduction in the strength of the C–O bond in comparison with free carbon monoxide. This is shown by the lengthening of the C–O bond and the reduction in the infra-red stretching absorption frequency on complexation, although no direct thermochemical measurements of C–O bond energies in carbonyls have been made. A consequence of this is that CO bonded to a metal is very unlike a free CO molecule, the differences probably being much greater than in bonded and free alkyl radicals, for example.

The stability of the metal-carbon bond owes much to the presence of available *d*-orbitals on the metal. It is for this reason that stable carbonyls of the transition metals, but not of the non-transition metals, are readily formed.

5.3.2 Thermochemistry of Carbonyls

Skinner [10] has commented that few reliable thermochemical studies of carbonyls have been made. The technique of rotating bomb calorimetry (Section 2.5.3) is appropriate to metal carbonyls which tend to burn to ill-defined products. Reliable values are restricted to $Mn_2(CO)_{10}$, $Cr(CO)_6$, and $Mo(CO)_6$, the last two being obtained from thermal decomposition and iodination studies. Other carbonyls have been examined by static bomb calorimetry but the results should be treated with caution. The value obtained by static bomb calorimetry for $Cr(CO)_6$, for example, is in error by about 11 per cent.

Mean metal—CO bond dissociation energies, for those carbonyls which have been studied thermochemically, are listed in Table 5.5, Data refer to the gas-phase dissociation into ground-state metal and carbon monoxide

$$\frac{1}{n} M(CO)_n(g) = \frac{1}{n} M(g) + CO(g) \quad \bar{D}(M\text{—}C) = \Delta H$$

Table 5.5. Enthalpies of Formation and Mean Metal—carbon Bond Dissociation Energies in Metal Carbonyls. (Data are from references [3] and [10], together with $\Delta_f H^\ominus(CO, g) = -110{\cdot}5$ kJ mol^{-1} and the values in Table 5.3.)

Compound	$\Delta_f H^\ominus$ (gaseous)/ kJ mol^{-1} at 298 K	$\bar{D}$(M—C)/kJ mol^{-1} at 298 K
$Cr(CO)_6$	−909	107
$Fe(CO)_5$	−724	118
$Ni(CO)_4$	−600	147
$Mo(CO)_6$	−912	151
$W(CO)_6$	−882	178

The strongest bonds are formed by those transition metals towards the centre of the series, emphasizing the importance of back electron donation by *d*-electrons from metal to CO. Bonding is favoured in $Ni(CO)_4$ since the tetrahedral configuration of the nickel atom has 3*d*-electrons available in all four positions. The iron atom in the trigonal bipyramidal complex $Fe(CO)_5$ has 3*d*-electrons available in only four of the five positions, and would be expected to have poorer back-bonding compared with the nickel case. In accord with this viewpoint, a larger mean bond dissociation energy is found in $Ni(CO)_4$ than in $Fe(CO)_5$.

Another noticeable trend is the increase in bond energy down the group chromium, molybdenum, tungsten. The chemical stability of these carbonyls is often in the reverse direction with $W(CO)_6$ being the most reactive. This occurs for example in the photochemical production of alkene and alkyne complexes from the hexacarbonyls, in which it is thought that the breaking of a M—CO bond to form $M(CO)_5\cdot$ radicals is an important step. A similar situation arises in the first transition series where the thermal stability is in the order $Fe(CO)_5 > Cr(CO)_6 > Ni(CO)_4$ but the mean $\bar{D}$ values are in the reverse order, Ni > Cr > Fe. As we mentioned earlier, bonded and free CO are considerably different from each other, and a better measure of bond strength may be the intrinsic bond dissociation energy, but these are difficult to calculate accurately. Calculations [11] for iron, chromium, and nickel produce $\bar{D}^*(Fe-CO) = 245$ kJ mol^{-1}, $\bar{D}^*(Cr-CO) = 225$ kJ mol^{-1}, and $\bar{D}^*(Ni-CO) = 191$ kJ mol^{-1} which are in the reverse order to the ground state bond dissociation energies. Further considerations which may influence the relation between stability and bond energy are:

(1) kinetic factors, and

(2) the difference between the first bond dissociation energy, D_1, which seems to be important in the decomposition of carbonyls, and the mean value, $\bar{D}$. We saw when dealing with stepwise bond dissociation energies (p. 108) that these two quantities can be considerably different.

The only binuclear carbonyl for which thermochemical data are at present available is $Mn_2(CO)_{10}$ [$\Delta_f H^{\ominus}(g) = -1615$ kJ mol^{-1}]. Electron impact studies [12] indicate a value of $\approx$79 kJ mol^{-1} for $D(Mn-Mn)$, leading to $\bar{D}(Mn-CO) = 99$ kJ mol^{-1}. A problem of how to apportion the total bond dissociation energy between the two types of metal–carbon bond exists for tricarbonylcyclopentadienylmanganese (0), $(C_5H_5)Mn(CO)_3$, [$\Delta_f H^{\ominus}(s) = -530$ kJ mol^{-1}]. We shall not deal in this book with carbonyl halides but thermochemical values for some of these are available [3, 10].

5.3.3 Thermochemistry of Carbonyls of Platinum-group Metals

The transition metals ruthenium, rhodium, palladium, osmium, iridium and platinum are often considered as a separate group because of fairly close similarities in their physical and chemical properties. They do not form simple carbonyls but some thermochemical information is available on rhodium and iridium mixed complexes, some of which show intriguing properties, although the CO group plays a passive role in these.

The complex chlorocarbonylbis (triphenylphosphine) iridium (I), known as Vaska's compound, shows an ability to catalyse a number of organic reactions, such as olefin hydrolysis, and an ability in solution to take up,

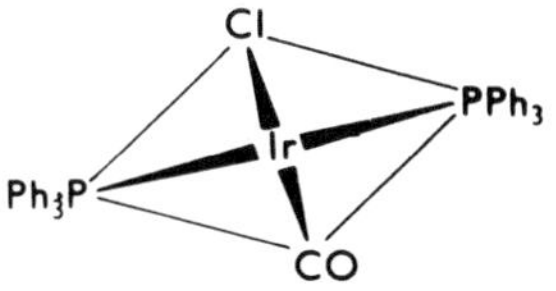

reversibly, gaseous oxygen which is released on reduction of O_2 pressure. In this latter respect, its action resembles that of haemoglobin which is the oxygen carrier in mammalian blood. The uptake of O_2 in solution in chlorobenzene ($\Delta H = -71$ kJ mol^{-1}) is thermochemically similar to that found ($\Delta H = -54$ to -75 kJ mol^{-1}) for the oxygenation of haemoglobin and other haemoproteins [13].

The iridium complex and its less reactive rhodium analogue, will also bond to SO_2, H_2, fluorinated alkenes and many other molecules, and the resulting adducts can often be obtained in the solid state. The structures [14] of two of these, the SO_2 and O_2 adducts are shown in Fig. 5.5. It is interesting

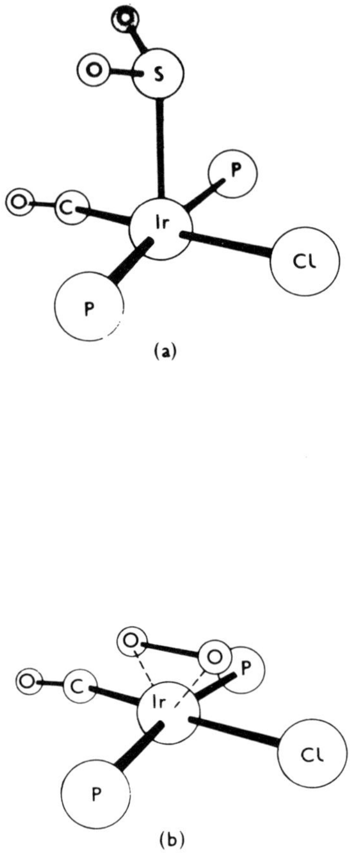

FIG. 5.5. Structure of (a) the SO_2 complex and (b) the O_2 complex of $IrCl(CO)(PPh_3)_2$

that, whereas SO_2 is bonded through sulphur in the expected way, albeit by an unusually long Ir–S bond, O_2 is bonded so that the O–O axis is at right-angles to the Ir–O bond.

These differences show up thermochemically in the data in Table 5.6 for the removal of gaseous ligand from the solid adducts. Values of ΔH for the SO_2 complexes, which may roughly be equated with $D(M{-}SO_2)$, are small, indicating weak bonds. This is also suggested by the length of the Ir–S bond, 0·249 nm, compared with about 0·207 nm in other SO_2 complexes.

The removal of gaseous oxygen from the iridium complex shows a strikingly different behaviour in that it is an exothermic process. This

Table 5.6. Enthalpy Changes [15] for the Removal of Gaseous Ligands from Crystalline Ir and Rh Complexes

Complex	Ligand removed	ΔH/kJ mol^{-1} at 298 K
$IrCl(CO)(PPh_3)_2SO_2$	SO_2	35
$RhCl(CO)(PPh_3)_2SO_2$	SO_2	33
$IrCl(CO)(PPh_3)_2O_2$	O_2	97

suggests that the O_2 molecule is drastically different in the complex compared with the free state. If we consider the removal of O_2 as occurring in two steps,

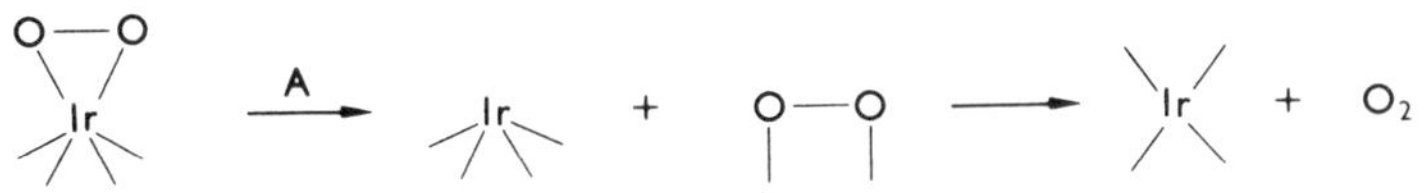

A the breaking of two Ir–O bonds, followed by *B*, the formation of molecular O_2 and the rearrangement of the remaining complex, then the enthalpy of step *B* may be approximated to the difference in bond energy terms E(O–O) in the oxygen molecule [16] (495 kJ mol^{-1}) and in the peroxide radical (146 kJ mol^{-1}). $\Delta H(B)$ becomes -349 kJ mol^{-1} and hence $\Delta H(A)$ = 349–92 = 257 kJ mol^{-1}, or $\bar{D}$(Ir–O) = 129 kJ mol^{-1}. This calculation neglects the rearrangement of the remaining Ir-ligand bonds, but information on these bond strengths is not available.

This Ir–O bond energy lends support to a peroxide formulation of the complex rather than π-type Ir–O_2 bonding, since the latter would presumably require much less O_2 reorganization energy on removal. It seems likely, however, that the Ir–O_2 interaction may be much more complex than envisaged in the preceding simple bond energy calculation.

5.4 Alkene and Alkyne Complexes: Two-electron Ligands

Transition elements showing marked tendencies to form bonds to alkenes and alkynes include Pt(0), Cu(I), Ag(I), and Hg(II), all of which have a d^{10} electronic structure. Thermodynamic data are at present confined almost entirely to platinum and silver, those for the latter referring to species in aqueous and ethyleneglycol solution. The 18-electron rule is again useful for rationalizing the formulae of alkene and alkyne complexes, particularly when other ligands are also bonded.

Alkynes form a great variety of complexes, in only a few of which the alkyne is considered to be a two-electron ligand. In these cases, the bonding resembles the metal–alkene bond discussed below. The reader is referred to Chapter 8 of Green's book [2] for a full survey of metal-alkyne complexes.

5.4.1 Bonding in Alkene Complexes

Alkenes resemble carbon monoxide in their ability to coordinate readily to transition metals which have available *d*-electrons. In the case of alkenes, the bond is at right-angles to the C–C axis. In the simplest alkene, ethylene, the C–C interaction consists of a σ-bond between two carbon sp^2 hybrid orbitals, and a π-bond, containing two electrons, between the two unhybridized $2p$ orbitals. Alkynes have one more C–C π-bond, but the mode of bonding to metals should closely resemble that for ethylene.

The metal-ethylene bond is considered to be a donation of two electrons from the bonding π-orbital of ethylene to a suitable σ-orbital on the metal (Fig. 5.6a). Further bonding occurs in the other direction with the donation

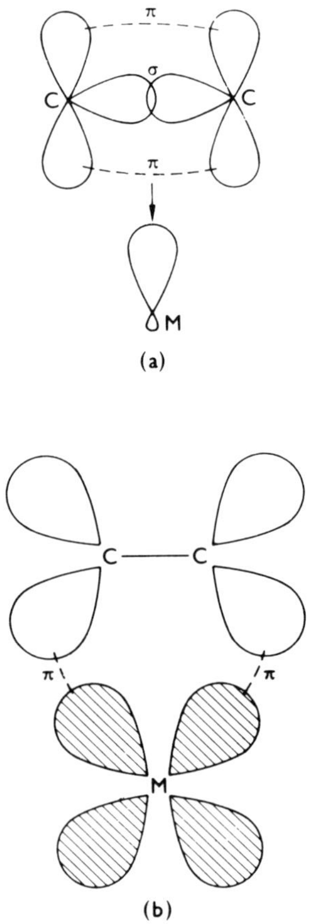

FIG. 5.6. Bonding in metal alkenes: (a) ligand-metal electron donation, (b) metal-ligand electron donation.

of electrons from filled $d\pi$-orbitals on the metal to the empty C–C $p\pi^*$ anti-bonding orbitals on ethylene (Fig. 5.6b). As in the case of carbon monoxide, this back-bonding reduces the accumulation of negative charge on the metal and leads to considerable bond strengthening over that possible with either type of electron donation alone. The fact that strong bonding occurs for those metals with a d^{10} electronic structure emphasizes that filled d-orbitals are required to form metal alkenes, although the relative importance of electron donation and back-bonding are not known.

5.4.2 Thermochemistry of Metal-alkene and Alkyne Complexes

A consequence of metal-ligand bonding in these systems is the removal of electronic charge from a bonding π-orbital on the ligand to an antibonding π^*-orbital on the ligand, with the result that the C=C bond should be weakened in comparison with the free ligand. Although no thermodynamic evidence is available to show this, the infra-red stretching 'frequencies' of the C=C bonds in alkenes are lowered by $\approx 100\ cm^{-1}$ on complexation. A lower frequency may qualitatively be linked to a weaker bond.

Replacement of hydrogen in the ligand by fluorine would be expected to reduce the donor properties of the alkene owing to the high electronegativity of fluorine. In the bonding of C_2F_4, for example, the back-bonding from metal to ligand should predominate over alkene-metal electron donation. As is shown below, this still allows a strong metal-ligand bond to be formed.

Bond dissociation energies

Metal-alkene and alkyne bonds are difficult to study thermochemically. Up to the present time, enthalpy values are available only for species in the solid state, from which bond dissociation energies may be obtained after several assumptions have been made.

Mortimer *et al.* [17] have studied the thermal properties of the complexes $PtXMeA_2L$ [where X = Cl or Br, A = $AsMe_3$ or $AsMe_2Ph$, and L = tetrafluoroethylene (C_2F_4) or hexafluorobut-2-yne (C_4F_6)] which lose the ligand L on heating according to the reaction

$$PtXMeA_2L(s) = trans\text{-}PtXMeA_2(s) + L(g)$$

Enthalpy changes for these reactions are shown in Table 5.7. On the assumption that the enthalpy changes will not be greatly different in the gas phase, values of ΔH may be equated to D(Pt–C). A more realistic estimate of the strength of the Pt–C bond might be made if the extent of the rearrangement of C_2F_4 and C_4F_6 on release from the complex could be made. Only for $PtClMe(AsMe_3)_2C_4F_6$ is structural information available. This shows that the C–C bond length in the complex (0·132 nm) is close to that in ethylene

Table 5.7. Enthalpy Changes for the Removal of Gaseous Ligand L from Crystalline $PtXMeA_2L$ Complexes to Give Crystalline *trans*-$PtXMeA_2$. (From Mortimer *et al.* [17])

Complex			ΔH/kJ mol^{-1}
A	X	L	
$AsMe_3$	Cl	C_2F_4	51·6
$AsMe_3$	Br	C_2F_4	48·2
$AsMe_3$	Cl	C_4F_6	68·6
$AsMe_3$	Br	C_4F_6	61·0
$AsMe_2Ph$	Cl	C_4F_6	73·4
$AsMe_2Ph$	Br	C_4F_6	80·3

(0·134 nm), but that free C_4F_6 has a C–C bond length of 0·122 nm, similar to that in free acetylene (0·120 nm). On the basis of these similarities, it can be estimated that the C–C bond strength will increase, as soon as dissociation from platinum occurs, by an amount $E(C\equiv C) - E(C=C)$, the bond energy terms in acetylene and ethylene, respectively. Cox and Pilcher [3] give these two values as 767 kJ mol^{-1} and 557 kJ mol^{-1}, respectively, the difference being 210 kJ mol^{-1}. Bond energy terms give this enthalpy directly since they are independent of the other bonds in the molecules. A similar figure could be calculated from bond dissociation energies after making allowance for the different numbers of C–H bonds in the two molecules.

The intrinsic bond dissociation energy of the Pt–C bonds is greater than the measured ΔH by 210 kJ mol^{-1}. This gives $D^*(Pt{-}C_4F_6) = 279$ kJ mol^{-1} indicating a moderately strong bond. It seems that the high electronegativity of the fluorine atoms, while influencing to some extent the strength of the Pt–C bond (a value of $\approx$300 kJ mol^{-1} might be expected for the unsubstituted ligand by analogy with D(Pd-allyl) in Section 5.5.2) still allows a substantial Pt–C bond to be formed.

Complexes in solution

Another series of thermodynamic measurements [18–20] concerns alkene and alkyne complexes of silver nitrate in solution in water or ethyleneglycol. It is not possible to relate the solution processes to bond dissociation energies in the gas-phase owing to the lack of data. This illustrates a frequent difficulty in the attempt to relate thermodynamics and bonding. For the solution work, we must be content to discuss trends.

Measurements relate to reactions of the type

$$AgNO_3 \cdot R(soln.) = AgNO_3(soln.) + R(soln.)$$

where R represents an alkene or alkyne molecule, and (soln.) refers either to water or ethyleneglycol. Two interesting experimental methods were used in these studies. For the species in ethyleneglycol solution, the procedure was a gas-liquid chromatographic one in which the retention time of alkenes on $AgNO_3$-ethyleneglycol/firebrick columns was measured. Comparison with columns without $AgNO_3$ allowed equilibrium constants for reactions of the above type to be calculated. Enthalpy and entropy changes were found from the variation of equilibrium constant with temperature (Section 2.5.1).

For the aqueous systems, the $AgNO_3$-alkene equilibrium constants were related to the increased solubility of the alkene (initially contained in CCl_4 solution) in aqueous $AgNO_3$ solution compared with aqueous KNO_3 solution. The latter was assumed not to form a complex with the alkene.

Table 5.8 contains thermodynamic data for a number of alkenes and alkynes, expressed in terms of the dissociation of ligand from the complex, with all species in solution.

In comparing the data for complexes in ethyleneglycol solution, it may be noted that the values of the entropy changes, ΔS, are almost constant whatever the ligand. Thus stability, as expressed by ΔG, follows closely the values of the enthalpy change, ΔH.

There is a general decrease in stability of $AgNO_3 \cdot R$ (decreasing ΔH) with increasing substitution of the double bond. Since alkenes are electron σ-donors and π-acceptors, substitution of the double bond by alkyl groups, (increased inductive effect) should increase the ligand's σ-donor properties but decrease its π-acceptor properties. It is not known at the present time which of the two bonding modes is the predominant bond strengthening factor, but the observed decrease in bond dissociation energy (ΔH) with increasing substitution favours the π-acceptor aspect. In the previous section, we saw that fluorinated alkenes and alkynes, with poor σ-donor properties, were still able to form strong Pt–C bonds. This also indicates a major π contribution to the bonding.

Although it is not possible to compare the values referring to aqueous solution for the cyclic alkenes and the alkynes, with those for complexes in ethyleneglycol solution, owing to the different solvent effects, an obvious trend amongst the cyclic alkenes is a decrease in stability in the order cyclopentene, cycloheptene, and cyclohexene. This parallels the order of decreasing ring strain in the uncomplexed cyclic alkenes. It seems likely that ring strain facilitates deformation of the alkene π-orbitals allowing the maximum metal-ligand bond strength through overlap with metal π-orbitals.

Table 5.8. Thermodynamic Data [18–20] **at** 298 K **for the Breaking of $AgNO_3$-alkene and -alkyne Bonds, with all Species in Solution and** $[AgNO_3] = 1$ mol dm^{-3}.

Alkenes in ethyleneglycol		ΔG/ kJ mol^{-1}	ΔH/ kJ mol^{-1}	ΔS/ J K^{-1} mol^{-1}
ethylene	$H_2C{=}CH_2$	7·1	14·6	25
		5·0	14·6	31
alkyl	$CH_3CH{=}CH_2$			
ethylenes	$C_2H_5CH{=}CH_2$	5·4	15·5	33
	n-$C_3H_7CH{=}CH_2$	4·7	15·1	34
	i-$C_3H_7CH{=}CH_2$	5·1	16·3	37
	cis-$CH_3CH{=}CHCH_3$	3·9	14·2	34
	trans-$CH_3CH{=}CHCH_3$	4·3	10·9	32
	cis-$C_2H_5CH{=}CHC_2H_5$	1·2	15·1	36
	trans-$C_2H_5CH{=}CHC_2H_5$	1·5	12·1	36
	$(CH_3)_2C{=}CHCH_3$	0·02	10·0	33
	$(CH_3)_2C{=}C(CH_3)_2$	2·7	7·9	36
	$C_2H_5(CH_3)C{=}CH_2$	3·7	14·6	36
di-olefin	$CH_2{=}CHCH{=}CH_2$	3·7	13·4	32
Cycloalkenes in aqueous solution				
	cyclopentene	−5·4	29·4	117
	cyclohexene	−9·9	23·3	111
	cycloheptene	−9·5	27·7	125
Alkynes in aqueous solution				
	$C_2H_5C{\equiv}CC_2H_5$		18·8	
	$CH_3C(CH_3)_2C{\equiv}CC(CH_3)_2CH_3$		20·9	

The metal-alkyne bond energies (ΔH) are of the same general order of magnitude as those for alkene complexes, as expected from the bonding similarities in the two cases.

5.5 Allyl Complexes: A Three-electron Ligand

The general name for hydrocarbon ligands which bond to transition metals by three-electron π-bonds is π-enyl. The simplest is the π-allyl group, π-$C_3H_5\cdot$, but it is also possible for this group to act as a one-electron σ-donor (as in Section 5.2), when it is written as σ-C_3H_5.

5.5.1 Bonding in π-allyl Complexes

The structure of π-allylpalladium (I) chloride dimer, $[\pi\text{-}C_3H_5PdCl]_2$, has been found by X-ray studies to be

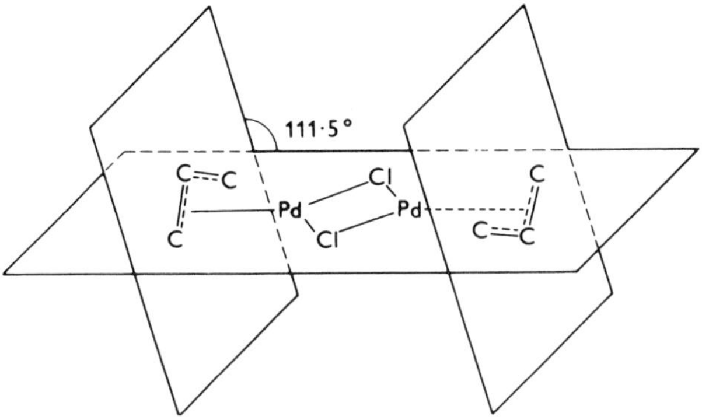

in which the three carbon atoms of each allyl group are in the form of an isosceles triangle. Three $p\pi$-orbitals on the carbon atoms form the three molecular orbitals shown at the top of Fig. 5.7. The first two of these are analogous to the bonding and antibonding π-orbitals in ethylene, whilst the third is characteristic of the allyl radical. The ligand C–H bonds are regarded as being unaffected by M–C bonding. Green [2] has discussed the possible metal orbitals which could result in overlap with the π-allyl molecular orbitals, and these are represented in Fig. 5.7 below the ligand orbital of appropriate symmetry. The first two columns of this figure would give a bonding situation similar to that of the metal-ethylene system, although in the allyl case the π-bond would contribute three electrons to the metal. Back metal-allyl bonding would arise from overlap of filled metal $d\pi$-orbitals with the antibonding orbitals of the allyl group. In the allyl case, the possibility shown in the third column of Fig. 5.7 also exists. It is impossible to say, on evidence presently available, which of the combinations shown in Fig. 5.7 are the most important contributions to the bonding.

We would expect a metal-allyl bond to have a strength at least as great as a metal-ethylene bond.

5.5.2 Thermochemistry of π-allyl Complexes

The π-allylpalladium chloride dimer is the only π-enyl complex which has been investigated thermochemically [21]. Differential scanning calorimetry, applied to the quantitative thermal reaction

$$[C_3H_5PdCl]_2(s) = 2Pd(s) + 2C_3H_5Cl(g)$$

leads to $\Delta_f H^\ominus(\text{complex, s}) = -107{\cdot}6\ \text{kJ mol}^{-1}$ and $\Delta_f H^\ominus(\text{complex, g}) = +18{\cdot}0\ \text{kJ mol}^{-1}$. The total of the bond dissociation energies in the complex is given by the enthalpy change for the process

$$[C_3H_5PdCl]_2(g) = 2Pd(g) + 2Cl(g) + 2C_3H_5(g)$$

$$\Delta H = 2\bar{D}(Pd{-}C_3H_5) + 4\bar{D}(Pd{-}Cl)$$

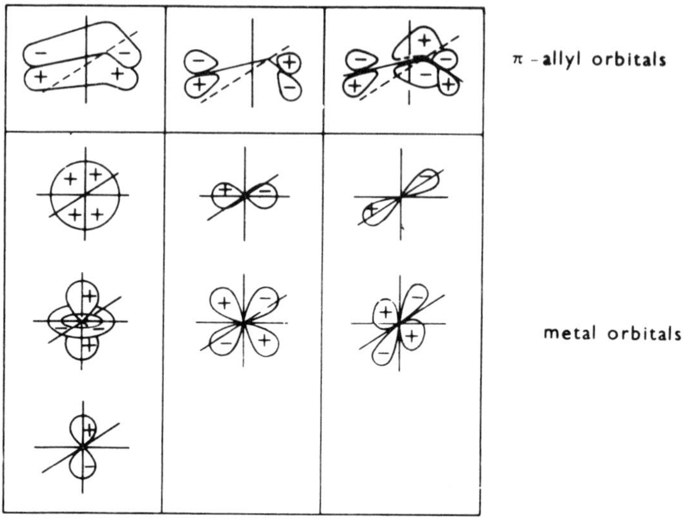

FIG. 5.7. π-allyl molecular orbitals, and metal orbitals suitable for overlap with them to form M–C bonds [2].

The enthalpies of formation of all species are known (Table 5.3) and hence $\Delta H = 1266$ kJ mol^{-1}.

A difficulty arises over the value of $\bar{D}$(Pd–Cl), but an estimate may be made as follows. The structure of crystalline $PdCl_2$ is chain-like, consisting of

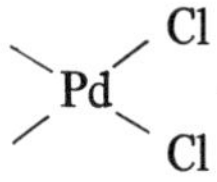

units. The enthalpy change for the reaction

$$PdCl_2(s) = Pd(g) + 2Cl(g) \quad \Delta H = 793{\cdot}1 \text{ kJ mol}^{-1}$$

will be similar to that required to break the Pd–Cl bonds in $PdCl_2$, although it will include a small but unknown contribution from the energy of interaction of the chains. On this basis we calculate a *maximum* value for D(Pd–Cl) of $\frac{1}{4} \times 793{\cdot}1 = 198$ kJ mol^{-1}. This in turn gives a minimum value for the Pd-allyl bond of $\bar{D}(Pd{-}C_3H_5) = 237$ kJ mol^{-1}. This does indeed indicate strong bonding between the π-allyl group and palladium.

5.6 π-Cyclopentadienyl and π-Benzene Complexes: Five- and Six-Electron Ligands

Hydrocarbon ligands which formally contribute five electrons to a metal-ligand bond are known as π-dienyl ligands. The best known is the cyclopentadienyl radical, $C_5H_5\cdot$; but other examples are to be found in 5-, 6-, and 7-membered cyclic and non-cyclic hydrocarbons. Cyclopentadiene may also bond as a σ-ligand and as the mainly ionic cyclopentadienides. Further details of these types may be found in Reference [2].

Six-electron ligands, of which benzene is the best known, are called π-arene ligands. Many other triene molecules act as six-electron ligands. Note that whereas cyclopentadienyl is bonded as a radical, benzene is bonded as a molecule.

Both these ligands frequently form bis(ligand)-transition metal complexes, such as bis(π-cyclopentadienyl)iron(0), $(\pi\text{-}C_5H_5)_2Fe$, and bis(π-benzene)-chromium(0), $(\pi\text{-}C_6H_6)_2Cr$. The former is also known as ferrocene, and the general name for this type of complex is 'metallocene'.

5.6.1 Bonding in 5- and 6-Electron Ligand Complexes

The π-orbitals in cyclopentadienyl and benzene are shown in Fig. 5.8.

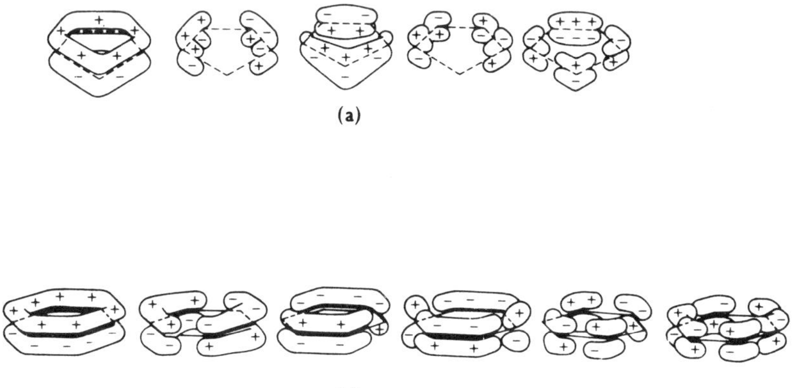

FIG. 5.8. *pπ*-molecular orbitals in (a) the cyclopentadienyl radical and in (b) benzene [2].

There are, of course, five of these for cyclopentadienyl and six for benzene. The C—H bonds are regarded as being unaffected by M—C bonding. In the complexes ferrocene and bis(benzene)chromium, the metals are 'sandwiched'

between the ligands. The problem of which metal orbitals are suitable for overlap with the ligand π-orbitals is one of group theory since symmetry considerations are of the utmost importance. This is outside the scope of the present text, and the reader is referred to the excellent treatment by Green [2]. Briefly, the five electrons in the π-orbitals of cyclopentadienyl or the six electrons of benzene may be donated to the metal to form a bond. Further bonding may occur by back donation from filled metal π-orbitals to the ligand.

Dienyl and arene ligands tend to form bis complexes, MR_2, from which it follows that very few can have eighteen valence electrons. Nevertheless, the 18-electron structure leads to particularly stable complexes. The rule is still useful in dealing with complexes of mixed ligands.

5.6.2 Thermochemistry of π-dienyl and π-arene Complexes

Mean metal-ligand bond dissociation energies may be calculated from the enthalpies of reactions of the type

$$\frac{1}{n}MR_n(g) = \frac{1}{n}M(g) + R(g) \quad \bar{D}(M\text{—}R) = \Delta H$$

when all the enthalpies of formation are known. A number of complexes have been studied by combustion calorimetry to yield the enthalpies of formation given in Table 5.9. The ligands mesitylene (1, 3, 5-trimethylbenzene) and *pseudo*cumene (1, 2, 4-trimethylbenzene) behave like benzene as six-electron ligands. For benzene and the substituted benzenes, which are stable molecules, the enthalpies of formation are well known. The value for the cyclopentadienyl radical, however, may be uncertain by up to ±40 kJ mol^{-1}.

The mean metal-ligand bond dissociation energies are collected in Table 5.9. The moderately high values parallel the thermal stability of the

Table 5.9. Standard Enthalpies of Formation and Mean Metal-ligand Bond Dissociation Energies in Gaseous Cyclopentadienyl and Benzene Complexes. (Data are from References [3] and [7])

Complex	$\Delta_f H^\ominus$/kJ mol^{-1}	$\bar{D}$(M—C)/kJ mol^{-1}
$(\pi\text{-}C_5H_5)_2Mg$	136·4	195
$(\pi\text{-}C_5H_5)_2Fe$	214·6	291
$(\pi\text{-}C_5H_5)_2Ni$	336·0	237
$(\pi\text{-}C_6H_6)_2V$	95·0	293
$(\pi\text{-}C_6H_6)_2Cr$	230·1	166
$(\pi\text{-}C_6H_6)_2Mo$	400·1	212
(π-1, 3, 5-trimethylbenzene)$_2$Cr	18·0	173
(π-1, 2, 4-trimethylbenzene)$_2$Cr	26·4	171
Ligands		
$\pi\text{-}C_5H_5\cdot$	190	
$\pi\text{-}C_6H_6$	82·9	
π-1, 3, 5-trimethylbenzene	−16·1	
π-1, 2, 4-trimethylbenzene	−13·9	

species, many of which are stable up to ≈600 K. A number of them can be sublimed in vacuum without thermal decomposition.

The values for ferrocene and nickelocene, $(\pi\text{-}C_5H_5)_2Fe$ and $(\pi\text{-}C_5H_5)_2Ni$, suggest appreciably weaker bonding in the latter. Ferrocene has an eighteen-electron structure, ten from the two ligands and eight from Fe(0). This, as mentioned earlier, tends to lead to a stable molecule in which all bonding orbitals are filled. In nickelocene, there are two additional electrons which must be placed in an antibonding orbital, so weakening the bond. It is interesting that X-ray diffraction studies show a longer M–C bond in nickelocene by about 0·015 nm compared with ferrocene.

Bond dissociation energy data are not available for cyclopentadienyl and benzene complexes of the same metal, but trends along the transition element series indicate a weakening bond for both types of complex. Since an early member of the transition series, $(\pi\text{-}C_6H_6)_2V$, has a mean bond dissociation energy close to that of a later cyclopentadienyl complex, $(\pi\text{-}C_5H_5)_2Fe$, it appears that the cyclopentadienyl ligand is the more strongly bonded.

Data for the chromium complexes with benzene and substituted benzenes, show that alkyl substitution has a negligibly small effect on the bond dissociation energies.

REFERENCES

[1] Rochow, E. G., *Organometallic Chemistry*, Chapman and Hall, London (1964).
[2] Coates, G. E., Green, M. L. H., and Wade, K., *Organometallic Compounds*, vols 1 and 2, Methuen, London (1967–68).
[3] Cox, J. D., and Pilcher, G., *Thermochemistry of Organic and Organometallic Compounds*, Academic Press, London (1970).
[4] Carson, A. S., Laye, P. G., Spencer, J. A., and Steele, W. V., *J. chem. Thermodynamics*, 2, 659 (1970).
[5] National Bureau of Standards, Technical Notes 270–3, 4, Washington D.C. (1968–69).
[6] Kerr, J. A., *Chem. Rev.* **66**, 465 (1966).
[7] Skinner, H. A., *Adv. Organometallic Chem.* 2, 29 (1964).
[8] Ashcroft, S. J., and Mortimer, C. T., *J. chem. Soc. (A)*, 930 (1966).
[9] Braterman, P. S., and Cross, R. J., *J. chem. Soc. Dalton Trans.* 657 (1972).
[10] Connor, J. A., Skinner, H. A., and Virmani, Y., *J. chem. Soc. Faraday Trans. I.* 1754 (1972).
[11] Skinner, H. A., and Sumner, F., *J. inorg. nucl. Chem.* **4**, 245 (1957).
[12] Bidinosti, D. R., and McIntyre, N. S., *Chem. Comm.* 555 (1966).
[13] Vaska, L., Chen, L. S., and Senoff, C. V., *Science*, **174**, 587 (1971).
[14] LaPlaca, S. J., and Ibers, J. A., *J. amer. chem. Soc.* **87**, 2581 (1965): *Inorg. Chem.* **5**, 405 (1966).
[15] Ashcroft, S. J., and Mortimer, C. T., *J. organometallic Chem.* **24**, 783 (1970): *Inorg. Chem.* **10**, 1326 (1971).

[16] Cottrell, T. L., *The Strengths of Chemical Bonds*, Butterworths, London (1958).
[17] Mortimer, C. T., McNaughton, J. L., and Pudephatt, R. J. *J. chem. Soc. Dalton Trans.* 1265 (1972).
[18] Cvetanović, R. J., Duncan, F. J., Falconer, W. E., and Irwin, R. S., *J. am. chem. Soc.* **87**, 1827 (1965).
[19] Traynham, J. G., and Sehnert, M. F., *J. amer. chem. Soc.* **78**, 4024 (1956).
[20] Helmkamp, G. K., Carter, F. L., and Lucas, H. J., *J. am. chem. Soc.* **79**, 1306 (1957).
[21] Ashcroft, S. J., and Mortimer, C. T., *J. chem. Soc.* (*A*), 781 (1971).

SUGGESTIONS FOR FURTHER READING

Bonding in Organometallic Compounds

Cotton, F. A., and Wilkinson, G., *Advanced Inorganic Chemistry*, Chs 22 and 23, 3rd edn, Interscience, New York (1972).

Orgel, L. E., *An Introduction to Transition Metal Chemistry*, Chs 9 and 10, Methuen, London (1960).

Stability of Metal–carbon Bonds

Mortimer, C. T., *Reaction Heats and Bond Strengths*, Ch. 8, Pergamon, Oxford (1962).

References [2] and [7].

Alkyl and Aryl Complexes

Parshall, G. W., and Mrowca, J. J., *Adv. Organometallic Chem.* **7**, 157 (1968).

Carbon Monoxide Complexes

Abel, E. W., and Stone, F. G. A., *Q. Rev. chem. Soc.* **23**, 325 (1969).
24, 498 (1970).

(The second reference contains a comprehensive list of reviews of carbonyl complexes.)

Chapter 6

Inorganic Polymers

6.1 Introduction

In Chapter 3, we discussed some thermochemical properties of the simplest inorganic polymers, the catenated hydrides. The scope of inorganic polymers is considerably wider than this [1] and, although organic polymers are more well studied and commercially exploited, inorganic chemists are becoming increasingly interested in the properties of more complex systems. There are two reasons for this:

(1) The realization of the potential usefulness of inorganic polymers as replacements, often with better properties, for the organic analogues.

(2) The increased sophistication of structural tools, nmr in particular, which have enabled investigators to probe the detailed internal structures of polymeric systems.

These inorganic systems exhibit a wide range of structural type and molecular geometries. By means of an approximate classification, we can distinguish between *homopolymers*, having the same element throughout the 'backbone', and *heteropolymers.* The latter is by far the largest and most technologically important class and its members most frequently have oxygen, nitrogen or sulphur atoms with atoms of one or more other elements in the backbone. These two basic types can be further subdivided and some examples are listed in Table 6.1.

There are two main features that distinguish organic and inorganic polymer systems. These are discussed below.

6.1.1 Bonding in Inorganic Polymers

Classical organic chemistry is concerned mainly with the combination of carbon, hydrogen, nitrogen and oxygen by the use of single or multiple covalent bonds. The atoms that comprise this list are limited to a coordination number of four.

In contrast, the situation for inorganic molecules is sometimes more complicated. This is because the atoms that comprise inorganic molecules

Table 6.1. Examples of Important Inorganic Polymer Systems

Type	Class	Examples
Homopolymers	Linear	Polysulphides, S–S–S–S–S–S Silanes, $SiH_3(SiH_2)_nSiH_3$
Heteropolymers	Unbranched	Phosphonitrilic Cl Cl Cl Chlorides, P–N–P–N–P–N Cl Cl Cl Polydimethylsiloxanes, $[(CH_3)_2SiO]_n$
	Branched	Polyarsenous oxyfluorides, –As–O–As–O–As– F F F
	Cyclic	Cyclic siloxanes and phosphonitrilic compounds
	Fused Cyclic	Polymonophenylsiloxanes, 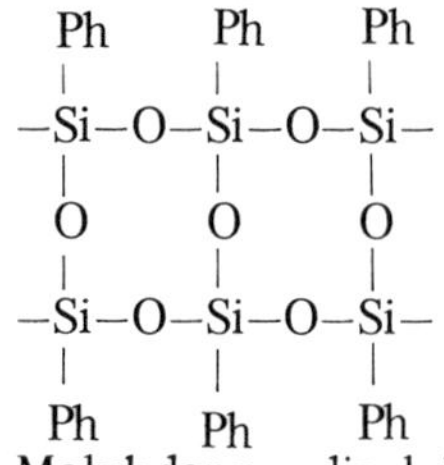
	Sheet	Molybdenum disulphide Boron nitride
	Networks	Silica, zeolites, glasses
π-complexes	Anionic	Copper (I) complex 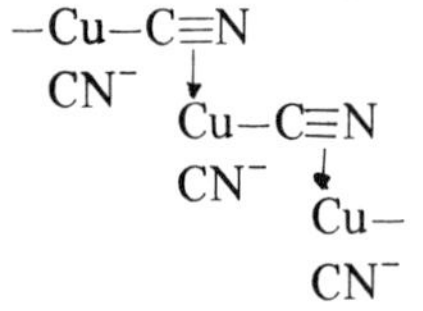
	Neutral	Copper (I) acetylide 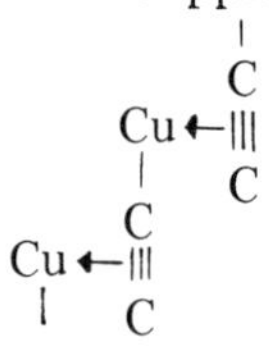

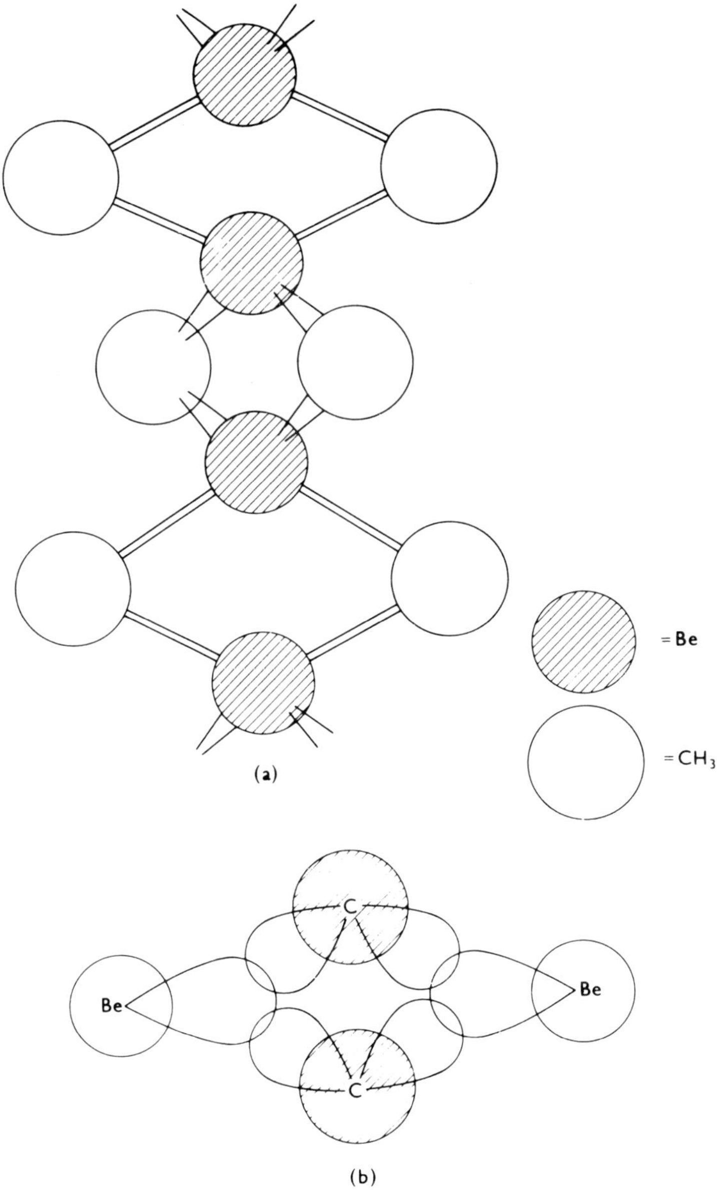

FIG. 6.1 (a) Chain structure of dimethylberyllium
(b) Orbital overlap of the beryllium and methyl orbitals

tend to use all the energetically available orbitals, sometimes regardless of whether there are enough valence electrons to fill the resultant covalent bonds. This is the case for the elements of Groups IIA, IIIB, and many transition metals. In each of these, the atom has less than one electron per available orbital. We have discussed previously (Section 3.5.2) the electron deficient bonding in boron hydrides; other examples are dimethyl beryllium, tetramethyl platinum, and several aluminium compounds. Figure 6.1 illustrates this type of bonding in polymeric systems.

Enhancement of organic polymer system properties has been achieved by means of the incorporation of π-complexes of the ferrocene and carborane type. Polymers of this type are, however, more correctly described as being mainly organic.

There is, also, a small number of completely π-bonded systems (Table 6.1). In many cases, however, the compounds are incompletely structurally characterized.

From our comments in Chapter 3 concerning the low stability of the catenated compounds of elements in the second and succeeding periods, it should be no surprise to learn that the heteronuclear polymers are more stable than the homonuclear analogues owing to the greater strengths of heteronuclear bonds.

For the heteropolymers of tin, antimony, tellurium and iodine we often observe coordination numbers in excess of four; other elements, notably silicon, phosphorus and sulphur, use d-orbitals in strengthening the existing σ-bonds whilst preserving a low coordination number. The latter group of polymers are, therefore, particularly stable.

The number of bonds formed by an atom in a polymer confers various physical properties. Of particular importance is the *functionality* of an atom – this being defined as the number of bonds formed by an atom to its adjacent bonded partners. We shall meet this term frequently in this chapter.

6.1.2 Equilibrium Control in Polymer Systems [2, 3]

In organic chemistry, there is considerable emphasis on the role of kinetics and the understanding of reaction mechanisms. The kinetics of the majority of inorganic reactions are less well understood owing to the greater lability of inorganic compounds. Indeed, the same products are often obtained as a result of quite different reaction mechanisms. This is one of the major differences between inorganic and organic preparative chemistry. The organic chemist makes great use of the relative inertness of the compounds which he is studying and, therefore, can make use of *kinetic* control – such as variation of reaction time or temperature – in order to obtain the desired products. With the exception of certain classes of reaction (such as the preparation of Co(III) and Pt(II) complexes), the inorganic chemist can only make use of *equilibrium* control.

This is particularly true for inorganic polymers. These compounds undergo rapid 'scrambling' reactions in which parts of the same or adjacent molecules exchange to give the equilibrium products.

Van Wazer and his co-workers [2] have made many studies of scrambling reactions; initially, these studies were directed towards an understanding of simple exchange reactions of monomeric inorganic compounds. More recently, many studies have been made of polymeric systems with the interesting result that there appear to be several similarities to the simpler exchange reactions of monomers. In particular, the scrambling reaction about a given central atom (to give the final, equilibrium products) seems to be controlled primarily by nearest neighbour interactions. Therefore, if we are considering the equilibration of a polymer, we shall expect that the more distant atoms of a polymer chain will have little influence on the scrambling about a given atom. This is very interesting since it is also known that the energies of covalent bonds are also primarily determined by first order interactions. Since the *breaking* and *making* of bonds is involved in scrambling reactions, it seems likely that the nearest neighbour rule may be controlled by enthalpy effects.

Let us consider the reaction

$$(\nu - i)\mathrm{QT}_\nu + (i)\mathrm{QZ}_\nu = \nu\mathrm{QT}_{(\nu - i)}\mathrm{Z}_i \tag{6.1}$$

where T and Z are exchangeable substituents and i and ν are integers. If there were completely random scrambling in (6.1), then the equilibrium constant could be calculated from statistical factors alone. In such a case we say that the reaction is fully *entropy-controlled.*

As examples of simple exchanging systems, let us consider the scrambling equilibria represented by

$$2\,\mathrm{MA_2B} \rightleftharpoons \mathrm{MA_3} + \mathrm{MAB_2} \tag{6.2}$$

$$2\,\mathrm{MAB_2} \rightleftharpoons \mathrm{MA_2B} + \mathrm{MB_3} \tag{6.3}$$

If the reactions are fully entropy-controlled we would calculate, purely on statistical grounds, that the equilibrium constant for each reaction should be 0·33. Let us examine two simple cases for monomeric substituents (applications to polymers will be discussed later in this chapter).

(a) M = PO; A = Cl; B = Br:
In this case, K(6·2) is 0·41 and K(6·3) is 0·34. The enthalpy change is very small for either reaction.

(b) M = P; A = Cl; B = $N(CH_3)_2$:
K(6·2) and K(6·3) are each equal to approximately 10^8. This is because the enthalpies of formation of $PCl_2N(CH_3)_2$ and $PCl[N(CH_3)_2]_2$ from PCl_3 and

$P[N(CH_3)_2]_3$ are 20·2 and 16·8 kJ mol^{-1} respectively. Therefore, (6.2) and (6.3) are each exothermic and strongly favour the products.

We have, in these reactions, examples of both entropy- and enthalpy-controlled reactions. They are, however, reactions of monomers. How can we apply our concepts to a discussion of polymers?

If, in (6.1), T is monofunctional and Z is bifunctional then the products of the scrambling reactions are seen to be fragments of a family of cross-linked polymers. The Z group could be, for example, a bridging atom or simply a bond to another Q atom. The process of equilibration from the reactants to the products still involves the making and breaking of bonds. We should be able to apply the concepts of entropy- and enthalpy-control to these polymeric systems although we must expect the mathematics of the problem to be more complicated.

In the remainder of this chapter we shall examine some typical examples of inorganic polymer systems. In some cases, it will be possible to discuss the experimental data in terms of the concepts we have introduced to describe scrambling equilibria. We shall also discuss the more conventional thermochemical data that have been obtained for some of the simpler polymer systems.

There remains a considerable amount of work to be done in this field. Until more reliable thermodynamic data are available, it will not be possible to give more than a qualitative view of the bonding in inorganic polymers. It is particularly unfortunate that little work has been done on the newer polymers or on the modified organic polymers, discussed previously in this section.

6.2 Examples of Inorganic Polymer Systems

6.2.1 Siloxane Polymers

These compounds comprise the largest class of commercially exploitable polymers. They may be prepared by the hydrolysis of alkyl-substituted chlorsilanes, R_nSiCl_{4-n}, followed by condensation. The degree of chlorination of the chlorsilanes dictates the extent of chain branching in the final polymer. We may summarize this as follows:

Reactant	*Hydrolysed Product*	*Condensed Product*	
R_3SiCl	R_3SiOH	R_3SiO-	(end group)
R_2SiCl_2	$R_2Si(OH)_2$	$-O-SiR_2-O-$	(middle)
$RSiCl_3$	$RSi(OH)_3$	$-O-SiR-O-$	(branch)
		|	
		O	
		|	

Only the tri-chlorinated silane yields the branch unit.

The product of the condensation reaction consists of a mixture of long and short chain polymers, together with cyclic compounds, (mainly the tetramer, $R_8Si_4O_4$) formed by the self-condensation of $SiR_2(OH)_2$.

Owing to the large Si–O bond energy (~480 kJ mol^{-1}), equilibration of products to a more homogeneous polymer proceeds slowly in the absence of a catalyst. Therefore the mixture is heated with dilute sulphuric acid, aqueous ferric chloride, or aqueous potassium hydroxide. In order to control the final molar mass, it is also common to add hexamethyldisiloxane, $Me_3SiOSiMe_3$, during the equlibration stage. This acts as a 'chain stopper' by providing Me_3SiO– and Me_3Si– end groups. For example, cyclic compounds undergo ring-opening reactions to give bifunctional units. After a small number of these units have joined together, the chain growth is terminated:

$$n(Me_2SiO)_4 \longrightarrow \overline{(SiMe_2O)_{4n}} \xrightarrow{Me_3SiOSiMe_3} Me_3SiO(SiMe_2O)_{4n}SiMe_3$$

In the absence of trichloromethylsilane in the original reaction mixture, fluid siloxanes are formed; when this silane is present the hydrolysis product, $MeSi(OH)_3$, produces resins owing to the formation of cross-linkages.

Equilibration favours chain, rather than cyclic polymers. This is related to the dependence of the bond energy, *D*(Si–O), on molecular geometry. The data in Table 6.2 show that *D*(Si–O) increases with chain length and is

Table 6.2. Si–O **and** Si–N **Bond Energies in some Silicon Compounds**

Compound	*D*(Si–O) /kJ mol^{-1}	Compound	*D*(Si–N) /kJ mol^{-1}
$Me_3SiOSiMe_3$	469	Me_3SiNEt	309
$Me_3SiOSi(Me)_2OSiMe_3$	482	$(Me_3Si)_2NH$	323
$Me_3SiO(SiMe_2O)_3SiMe_3$	494	$(Me_2SiNH)_4$	343
$(Me_2SiO)_4$	451		

smallest for the cyclic compound. The increase in *D*(Si–O) with chain length is probably due to increased electron delocalization and relief of ring strain. By comparison, values of *D*(Si–N) appear to indicate that the cyclic compound is the more stable. Unfortunately, data for the analogous chain compounds are not available.

Scrambling Reactions of Silicon Compounds [3]

The scrambling reactions of non-polymeric silicon compounds, as defined by (6.4), (6.5) and (6.6) (for non-coordinating solvents),

$$2SiXY_3 = SiX_2Y_2 + SiY_4 \qquad (6.4)$$

$$2SiX_2Y_2 = SiX_3Y + SiXY_3 \quad (6.5)$$

$$2SiX_3Y = SiX_4 + SiX_2Y_2 \quad (6.6)$$

are nearly always random, corresponding to almost thermoneutral reactions.

However, with X = –OR or $-NR_2$ and Y = halogen, there is non-random scrambling, as seen in Table 6.3. In general, those reactions with the larger

Table 6.3. Equilibrium Constants and Enthalpy Changes for the Scrambling Equilibria (6.4)–(6.6) in the Liquid Phase

X	Y	Temp/°C	K^a(6.4)	ΔH(6.4) /kJ mol^{-1}	K^a(6.5)	ΔH(6.5) /kJ mol^{-1}	K^a(6.6)	ΔH(6.6) /kJ mol^{-1}
OMe	OEt	150	0·417	−0·394	0·422	0·205	0·386	−0·10
Et	*n*–Pr	180	0·694	–	0·315	–	0·454	–
Me	Cl	350	–	–	0·025	–	0·013	–
MeO	Cl	120	0·094	+12·0	0·042	+7·7	0·009	+12·2
Me_2N	Cl	25	6×10^{-4}	+16·3	5×10^{-4}	+17·4	6×10^{-6}	+27·1

[a]Random values of *K*(6.4), *K*(6.5), *K*(6.6) are 0·375, 0·444, and 0·375 respectively.

enthalpy changes show the greatest deviation from randomness.

It is of interest to compare the behaviour of the monomeric systems with the closely related polymeric systems. The siloxanes described in Section 6.2.1 are not amenable to accurate measurements owing to their relative complexity and inertness. Studies have been made on related systems containing more labile groups. A good example is the reaction between tetradimethylsiloxane and SiY_2Me_2, where Y is a labile substituent such as Cl, –OMe, –SMe, or $-NMe_2$. Figure 6.2 illustrates the kinetic dependence of a typical reaction. Also in Fig. 6.2 is shown a reaction involving tridimethylsilazane, a nitrogen-containing compound. There is a clear difference in the equilibrium concentrations of end-units, Me_2SiYZ-, middle units, $Me_2Si\langle^{Z-}_{Z-}$, and the so-called *neso*-molecules, Me_2SiY_2. (Y and Z being mono- and di-functional groups, respectively.) From the results, equilibrium constants for equation (6.7) can be derived.

$$2 \text{ ends} = \textit{neso} + \text{middle} \quad (6.7)$$

The values are shown in Table 6.4. Where a comparison is possible, the deviations from randomness compare well with those which are analogous in the monomeric systems – thus confirming the nearest-neighbour hypothesis.

Within a related series, *K*(6.7) measures the stability of –Z–Si–Z– middle groups compared with Si–Z– end groups; larger values favouring the former. The low value for Z = NMe reflects the low value of *D*(Si–N) in linear polymers

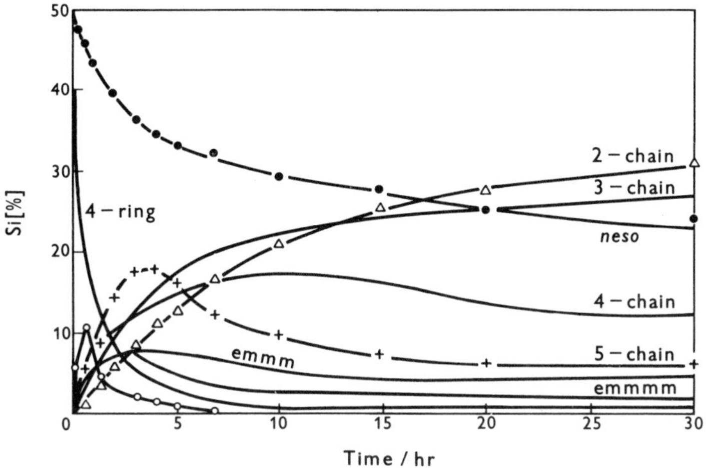

(a)

(a) Variation with time of the distribution of total silicon present in various molecules or building units during equilibration at 200°C of a mixture of one mole of dichloro-dimethyl-silane with $\frac{1}{4}$ mole of octamethylcyclotetrasiloxane using 0·5 wt % $AlCl_3$ as catalyst. To avoid confusion, experimental points are shown only for the *neso*-compound (•), the di-chain (△), the penta-chain (+), and the large molecules (big rings or long-chains) (○).

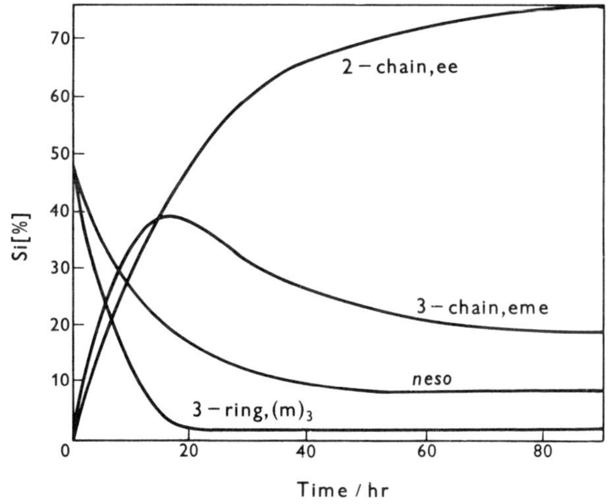

(b)

(b) Variation with time of the percentage of total silicon present in various molecules during equilibrium at 120°C of a mixture of one mole of dichloromethylsilane with $\frac{1}{3}$ mole of nonamethylcyclotrisilazane. The marks along the top of this figure denote the times at which the data were obtained.

FIG. 6.2 Kinetic dependence of the reaction between $SiCl_2Me_2$ and (a) $Si_4Me_8O_4$ or (b) $Si_3Me_9N_3$ in the liquid phase (from reference [2]).

expected (but not experimentally confirmed) from the data in Table 6.2. With Z=O or S, more nearly random values are obtained. The almost perfectly random value for Z=O and Y=MeO– suggests that equilibration of the more complex siloxanes should be random. This is thought to be true on the basis of other measurements [3].

Table 6.4. Equilibrium Constants in the Systems $SiMe_2X_2$ (A) and $(Me_2SiY)_n$ (B) for Reaction (6.7) in the Liquid Phase.

A	B	K^a(6.7)	Temp/°C
$SiMe_2Cl_2$	$(Me_2SiO)_4$	0·11	200
	$(Me_2SiS)_3^b$	0·13	200
	$(Me_2SiNMe)_3^b$	$2{\cdot}7 \times 10^{-3}$	120
$SiMe_2(MeO)_2$	$(Me_2SiO)_4$	0·24	200
	$(Me_2SiS)_3^b$	0·40	200
$SiMe_2(MeS)_2$	$(Me_2SiS)_3$	0·20	200
$SiMe_2(Me_2N)_2$	$(Me_2SiO)_4$	0·50	200

[a] Random value is 0·25.
[b] Ring formation is important in these systems.

The ring-chain equilibria (6.8) in the liquid phase have also been studied. At 120°C, with

$$\mathrm{T{-}SiMe_2{-}[Z{-}SiMe_2]_{i+r}{-}T} = \mathrm{[SiMe_2{-}Z]_r} + \mathrm{T{-}SiMe_2{-}[Z{-}SiMe_2]_i{-}T} \qquad (6.8)$$

T = Cl and $r = 3$, the equilibrium constant for (6.8) is 0·5 mol dm^{-3} for Z = O; 40 mol dm^{-3} for Z = NMe; and 2000 mol dm^{-3} for Z = S. These data clearly describe the increasing preference for the formation of cyclic molecules.

6.2.2 Phosphorus-nitrogen Polymers [4, 5]

The most important representatives in this group are the phosphonitrilic compounds, $(NPX_2)_n$, X being a monofunctional group such as chlorine. At one time it was hoped that they would be commercially useful but their instability towards moisture seriously impairs their widespread application.

The chloro derivatives are prepared by reacting ammonium chloride and phosphorus pentachloride in *sym*-tetrachloroethane. The product is a mixture containing small ring compounds, $(NPCl_2)_n$ mainly with $n = 3$ or 4, and short chain oligomers, $PCl_4(PNCl_2)_nCl$. On heating to 250–300°C, all of the lower members polymerize to yield a solid linear polymer with a molar mass of up to 200 000 g mol^{-1}. The atmospheric degradation of this polymer, referred to above, occurs with cross-linking owing to the ease of hydrolysis of the P–Cl linkage (Fig. 6.3).

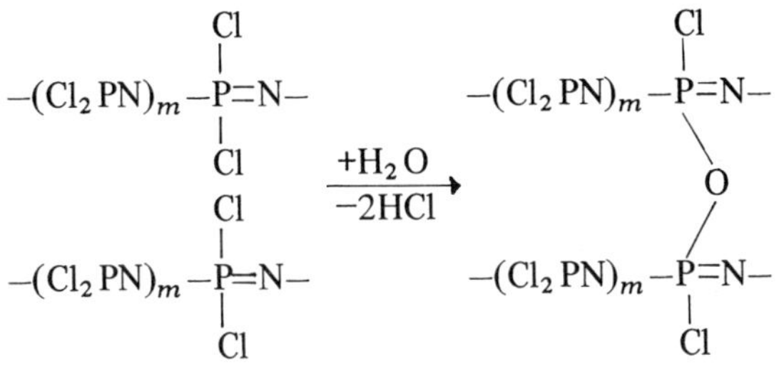

FIG. 6.3. Hydrolysis and cross-linking of a phosphonitrilic chloride.

The cyclic phosphonitrilics

The bonding in these compounds has received more attention than that in the linear polymers. The trimers, $(PNX_2)_3$, are almost planar but the higher members deviate from strict planarity and, in some cases, there is the possibility of configurational isomerism.

The π-bonding component is considered separately from the σ-bonding and a suitable scheme is shown in Fig. 6.4. A molecular orbital treatment [5]

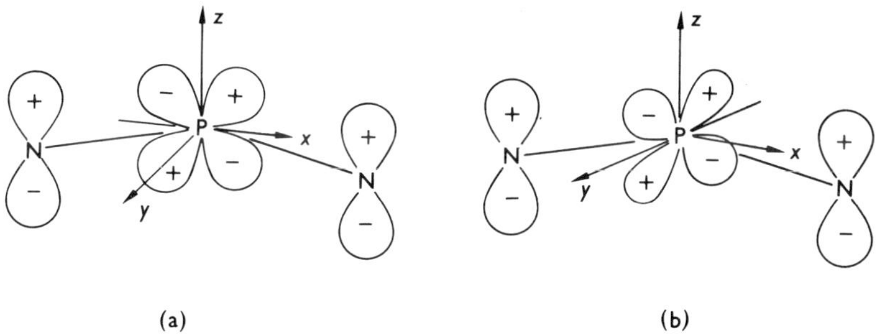

FIG. 6.4. Overlap of phosphorus d_π orbitals with nitrogen p_z: (a) d_{xz} (b) d_{yz} overlap.

involving the phosphorus d-orbitals and nitrogen p_z-orbitals has shown that:

(1) The phosphorus d_{xz} overlaps more efficiently than the d_{yz} with the nitrogen p_z orbitals (Fig. 6.4).

(2) Any even number of π-electrons completes a closed shell.

(3) π-electron energies increase as a function of ring size.

The above results are quite different from those obtained for conjugated cyclic carbon compounds. Simple π-electron theories of these compounds predict much more irregular behaviour in that closed shells (and large delocalization energies) only occur with $4n + 2$ π-electrons. In Fig. 6.5, we

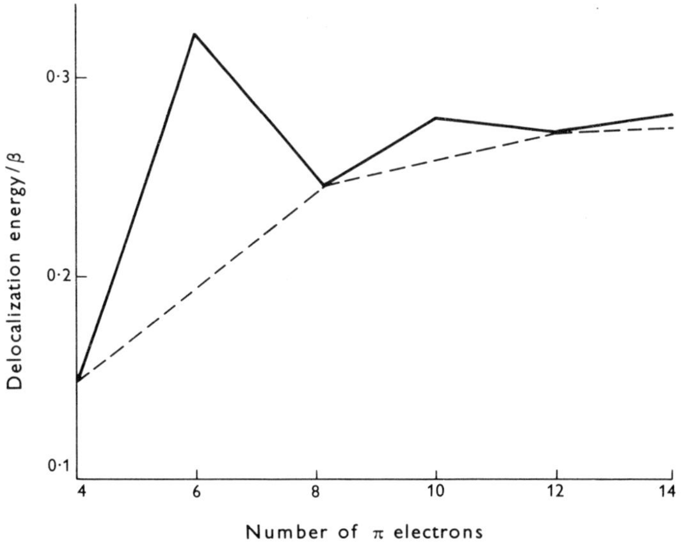

FIG. 6.5. Delocalization energies in planar $(AB)_n$ systems. Full lines refer to p_π–p_π (organic molecules) and broken lines to p_π–d_π, $(PNCl_2)_n$.

show the delocalization energies to be expected in cyclic $(AB)_n$ systems. The quantity β is an integral of the form

$$\beta = \int \psi_A H \psi_B \, dv$$

and this characterizes the strength of the A–B interaction (see Section 1.3).

We may estimate the delocalization energies from experimentally determined enthalpies of formation, which are listed in Table 6.5. From the

Table 6.5. Thermochemical Properties of $(NPCl_2)_n$

	$(NPCl_2)_3$	$(NPCl_2)_4$
$-\Delta_f H$ (s)/kJ mol^{-1}	812	1084
$[D(P–N) + D(P–X)]$ /kJ mol^{-1}	637	639
$\delta\,[D(P–N) + D(P–X)]$[a]/kJ mol^{-1}	39	41

[a]excess of measured $D(P–N) + D(P–X)$ over the sum of the single bond values.

results, we note that the experimentally determined values exceed the single bond energies by amounts that are in broad agreement with the above theoretical discussion. It is, however, difficult to proceed to a more detailed analysis of P—N bond energies. The main point which has emerged is a clear difference from the analogous carbon compounds, benzene and cyclooctatetraene.

Polymerization

The cyclic compounds polymerize on heating to about 300°C. The polymerization is catalysed by, for example, metals or organic oxygen-containing compounds. The mechanism is thought to involve the initial removal of a chloride ion, followed by the ring-opening attack of the cation on another cyclic molecule.

e.g.

$$P_3N_3Cl_6 \rightarrow P_3N_3Cl_5^+ + Cl^-$$

$$P_3N_3Cl_5^+ + P_3N_3Cl_6 \rightarrow P_6N_6Cl_{11}^+$$

The linear cation continues to grow by similar ring-opening reactions until the chains are terminated by end groups such as PCl_4^+, Cl^- or by impurities. Depolymerization begins at about 350°C and an equilibrium is set up between cyclic molecules at 600°C.

From the energies of activation for polymerization and depolymerization, the enthalpies of polymerization may be calculated and these are shown in Table 6.6. A large part of $\Delta^p H$, the enthalpy of polymerization, comes from

Table 6.6. Enthalpies of Polymerization, $\Delta^p H$, of Liquid Phosphonitrilic Chlorides

	$(PNCl_2)_3$	$(PNCl_2)_4$	$(PNCl_2)_5$	$(PNCl_2)_6$
$-\Delta^p H$, $[1/n\,(PNCl_2)_n, l]$ /kJ mol^{-1}	1·9	0·92	0·67	0·17

the increasing delocalization of the nitrogen lone pair, as evidenced by the large bond angle ($P\hat{N}P$ = 127°) in the nearly flat polymer. Similarly, the decreasing values of $\Delta^p H$ with increasing ring size are consistent with increasing electron delocalization in the rings.

The polymerization can be described by the scrambling reactions inherent in the above mechanism but no detailed analysis has been made.

6.2.3 Phosphorus- and Arsenic-oxygen Polymers

This group of compounds was important in establishing the theory of scrambling reactions. For example, P_2O_5 and $POCl_3$ undergo rapid scrambling reactions in the liquid phase to yield polyphosphoryl chlorides (Fig. 6.6a), whereas P_2O_5 and $PO(NMe_2)_3$ undergo non-random scrambling to yield polyphosphoryldimethylamides (Fig. 6.6b). In addition to the linear polymer,

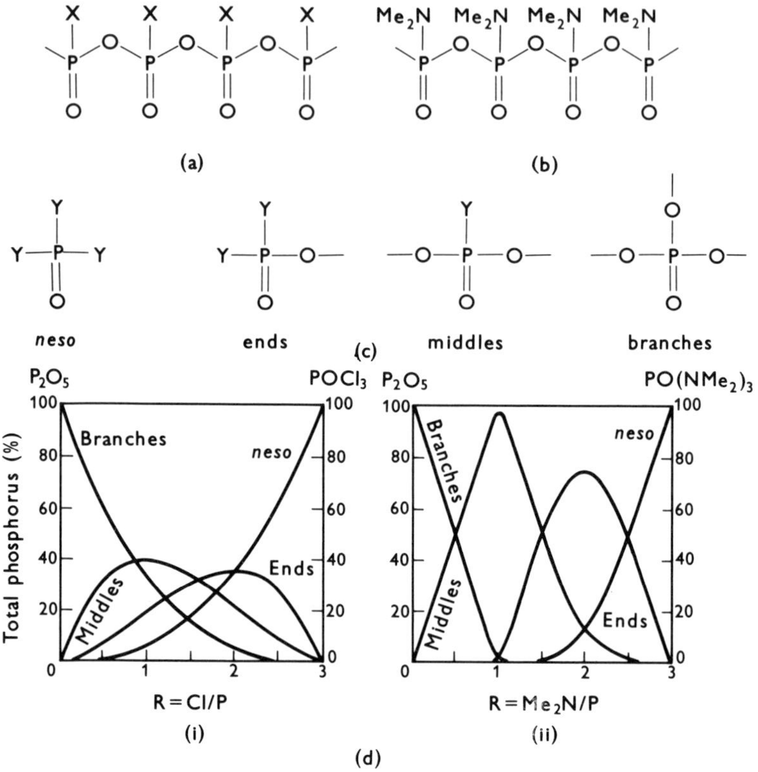

FIG. 6.6. Phosphorus-oxygen polymers (a) polyphosphoryl halides (b) polyphosphoryldimethylamides (c) building units (d) distribution of building units (i) in the phosphorus pentoxide-phosphorus oxychloride system (ii) in the phosphorus pentoxide-orthophosphoryldimethylamide system (from reference [1]).

structures with varying degrees of cross-linking occur in the equilibrated systems, and the building units required are illustrated in Fig. 6.6(c). Finally, in Fig. 6.6(d), we illustrate the distribution of the building units in each of the two systems. The strong, non-random, preference for middle units with $[Me_2N]/[P] = 1$ results in the polymer of this composition being composed entirely of middle units. This explains the experimental observation that this polymer has elastomeric properties.

Studies have also been made of the families of compounds derived from the systems MX_3/M_4O_{10}, in which M is phosphorus or arsenic, X is a monofunctional substituent (F, Cl, OMe and NMe_2) and Y is a difunctional substituent (O, NMe, and S).

From nmr data it has been possible to calculate equilibrium constants between the *neso* molecules, MX_3, and the various end, MX_2Y, and middle, MY_2X, units. The equilibria identified were (6.9) and (6.10) and the values of the equilibrium constants (in the liquid phase)

$$2 \text{ middles} = \text{end} + \text{branch} \quad (6.9)$$

$$2 \text{ ends} = \textit{neso} + \text{middle} \quad (6.10)$$

are shown in Table 6.7. As can be seen, the arsenic systems are nearly random,

Table 6.7. Equilibrium Constants for Equations (6.9) and (6.10) in the Liquid Phase

M	X	Y	$K(6.9)^a$	$K(6.10)^a$
PO	Cl	O	0·422	0·64
PO	NMe_2	O	6×10^{-4}	0·026
AsO	F	O	0·31	0·33
AsO	OMe	O	0·29	0·36
AsO	NMe_2	O	4·0	0·2

[a]Random value of $K(6.9)$ and $K(6.10)$ is 0·333.

with the exception of $X = NMe_2$. The values of the equilibrium constants suggest that, in this case, there is some degree of enthalpy control. The deviation from randomness is greater in the analogous phosphorus system and, in this case, the enthalpy changes of both (6.9) and (6.10) must be very endothermic. It is of interest to recall that a similar behaviour was noted for the scrambling between the monomeric compounds PCl_3 and $P(NMe_2)_3$, in Section 6.2.2. This is further justification for believing that a thorough understanding of the thermodynamic properties of monomeric molecules is essential before progress can be made in the more complicated field of polymers. More work along the lines established by Van Wazer and his co-workers, in addition to classical thermochemistry, is needed before a more complete view can be achieved.

REFERENCES

General

[1] Currell, B. R., and Frazer, M. J., 'Inorganic Polymers' in *R. Inst. Chem. Rev.* 2, 13 (1969).

Scrambling Reactions

[2] Van Wazer, J. R., and Moedritzer, K., *Angew. Chem. int. Ed.* **5**, 341 (1966).

[3] Lockhart, J. C., *Redistribution Reactions*, Academic Press, London, (1970).

Phosphonitrilics

[4] Paddock, N. L., *Q. Rev.* **18**, 168 (1964).

[5] Craig, D. P., *Theoretical Organic Chemistry*, Butterworths, London (1959).

Chapter 7
Aqueous Solutions and Non-Aqueous Solvents

7.1 Properties of Solvents

7.1.1 The Structures of Liquids [1, 2]

Many substances are capable of existing in the solid, liquid, or gaseous states under defined conditions of temperature, pressure, and volume. Only at one particular combination of pressure and temperature called the 'Triple Point' can a liquid phase be in equilibrium with both the solid and gaseous phases.

The degree of structural change on passing from solid to liquid is much smaller than that observed on passing from liquid to gas. This is inferred from the relative smallness of the changes in physical and thermodynamic properties in the former phase change. This similarity of solids and liquids is confirmed structurally by X-ray diffraction data, examples of which are shown in Fig. 7.1 for gaseous, liquid, and solid zinc. The existence of such diffraction patterns is evidence that the immediate environment of an atom or molecule in the liquid state is similar to that found in a solid.

Theoretical models of the structure of liquids are statistical in nature, owing to the large number of ways in which the atoms or molecules of a liquid may be arranged spatially. As a simplification, a number of lattice theories of the liquid state have been devised and the common feature of such models is that the proposed structures of liquids bear some resemblance to the extremely regular structure of the corresponding solid phase. Other theories, which may be called 'Distribution Function Theories', emphasize the process by which intermolecular forces determine the structure of a liquid. Although we shall be concerned mainly with intermolecular forces in liquids and solutions, it should be remembered that we do not regard either type of theory as being the more correct, since each is complementary to the other.

We shall now discuss the main categories of liquids and we shall also discuss the properties of water in a separate section owing to the special importance of this solvent in inorganic chemistry.

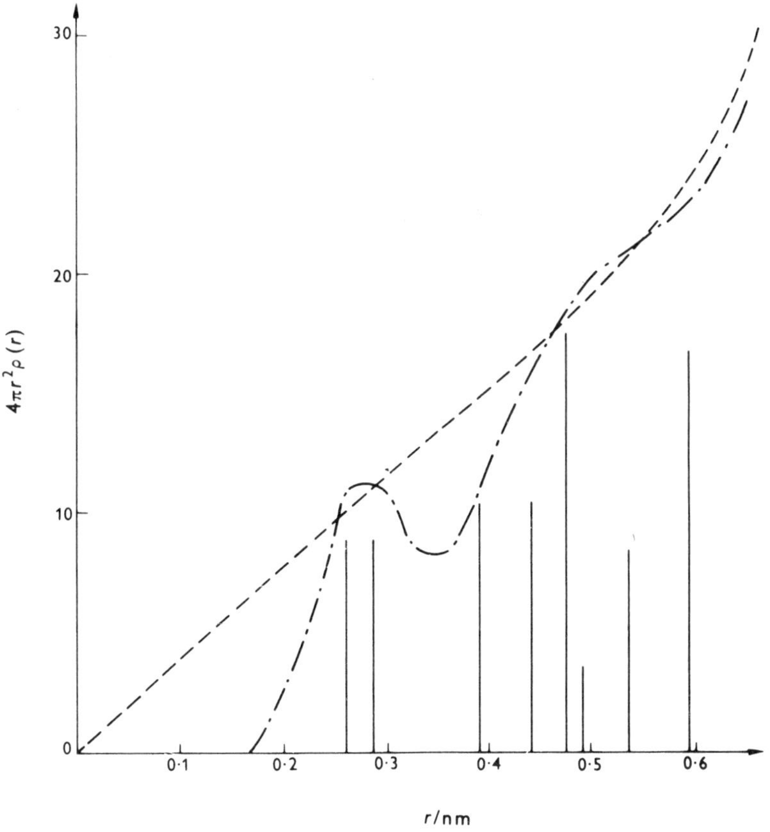

FIG. 7.1. Density distribution function ($\rho(r)$ = density of atoms at r/nm) for solid,———, liquid,— · —, and gaseous,– – – – –, zinc.

7.1.2 Noble Gases and Other Non-polar Molecules

The atoms in a noble gas are spherically symmetrical and the only attractive forces are 'dispersion forces' (also referred to as 'van der Waals' forces'). These have their origin in the attraction between the oscillating charge clouds of an atom and those of neighbouring atoms. This dipolar fluctuation induces a synchronized dipole in the neighbouring atoms and a nett attraction results. This is illustrated in Fig. 7.2. These attractive forces increase with the number of electrons in an atom and, also, with the number of atoms in a non-polar molecule.

The magnitude of an induced dipole can be estimated quantitatively from the polarizability, α, of an atom or molecule; α can be measured from the results of dielectric constant measurements. A theoretical treatment relates

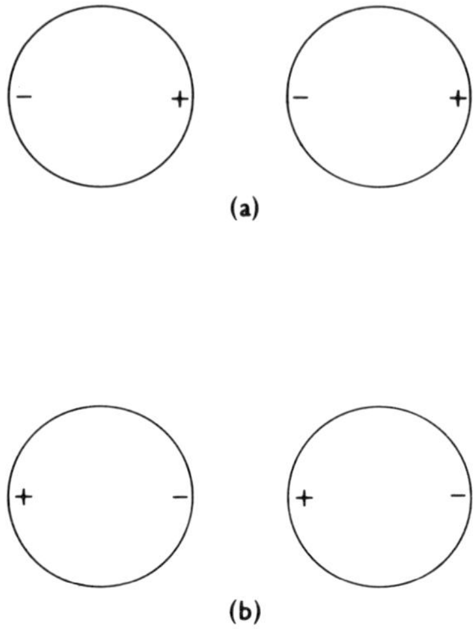

FIG. 7.2. Sequential representation (a) and (b) of the synchronized fluctuation of electronic charge in two adjacent atoms.

α to E_D, the cohesive energy between molecules, by the equation

$$E_D = -\alpha^2/r^6$$

r is the 'van der Waals' ' radius which may be thought of as one half of the closest distance of approach between two atoms or molecules. Note that the only contributions to E_D are dispersion forces which are only of importance in non-polar molecules or noble gas atoms in the liquid phase.

The enthalpy of vaporization, $\Delta_l^g H$, of a liquid should be a measure of the cohesive energy between the molecules. In Fig. 7.3 are shown the values of $\Delta_l^g H$ as a function of polarizability for a number of homonuclear and heteronuclear molecules. Although the homonuclear molecules and noble gases appear to conform to the polarizability arguments presented above, the heteronuclear molecules show gross divergences from the behaviour expected from dispersion forces alone.

7.1.3 Heteronuclear Molecules and Hydrogen Bonding [3]

In addition to the relatively weak dispersion forces mentioned above, there are also the stronger *permanent* dipole–dipole forces that exist between heteronuclear molecules. These forces are responsible for the anomalously high values of $\Delta_l^g H$ in Fig. 7.3. A special case of dipole–dipole bonding is hydrogen-bonding. This occurs when hydrogen, an atom of low electronegativity, is near to an atom of high electronegativity – particularly nitrogen, oxygen or fluorine. H-bonding can occur within a molecule (intramolecular) or between molecules (intermolecular). It is the latter

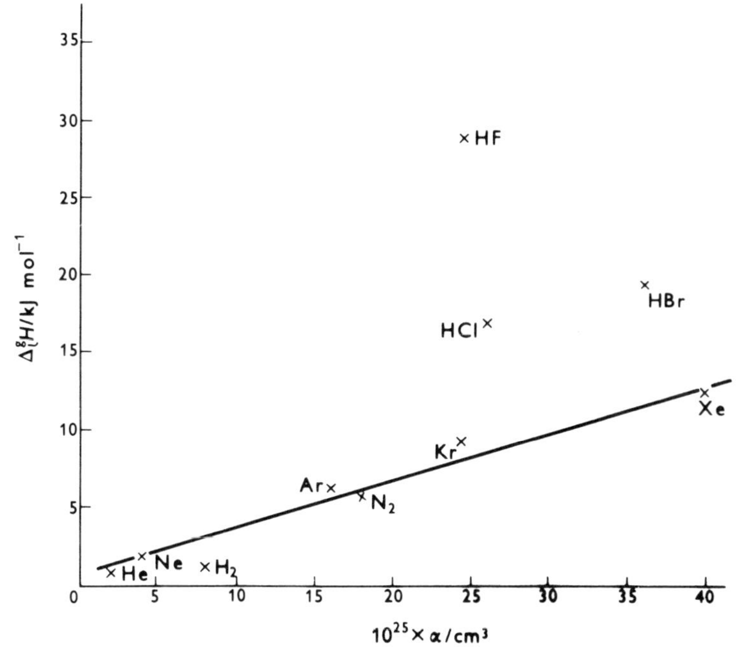

FIG. 7.3. Enthalpies of vaporization of some atoms and molecules as a function of polarizability, α.

which is the more important with respect to interactions in liquids. There are at least four contributions to the energy of a H-bond of the type A–H . . . B where B is an electronegative atom of a molecule which is H-bonded to A–H. The contributions are: (1) electrostatic (2) delocalization (3) dispersion, and (4) repulsive interactions. The electrostatic interaction arises because of an attraction between positive charge on the H atom and the lone pair of the

Table 7.1. Estimated Energy Contributions, E, to an Individual Hydrogen Bond in Ice

		E/kJ mol^{-1}
Electrostatic (coulombic)		−25·1
Delocalization (charge transfer)		−33·5
Dispersion		−12·5
Repulsive Overlap		+35·1
	Total, E_T	−36·0
One half of $\Delta_s^g H^\ominus$(ice)/kJ mol^{-1}		−25·5

donor atom, B. As the H-bond shortens, covalent bonding contributions of the type A–H . . . B increase in importance together with dispersion forces (remember the r^6 dependence). At shorter distances repulsive forces, a consequence of the Pauli Principle, destabilize the H-bond.

Some estimates of the importance of (1) to (4) were made originally by Coulson for H-bonding in ice. These are shown in Table 7.1. The total stabilization energy is similar to one half of the enthalpy of sublimation of ice, $\Delta_s^g H^\ominus$(ice). It is debatable whether the latter is strictly a valid measure of the H-bond energy in ice, but it may be a good approximation since *two* H-bonds need to be broken before a molecule of water can be put into the gas phase.

Molecular orbital descriptions of H-bonds have been attempted in recent years. For example, the Extended Hückel Theory was applied to the

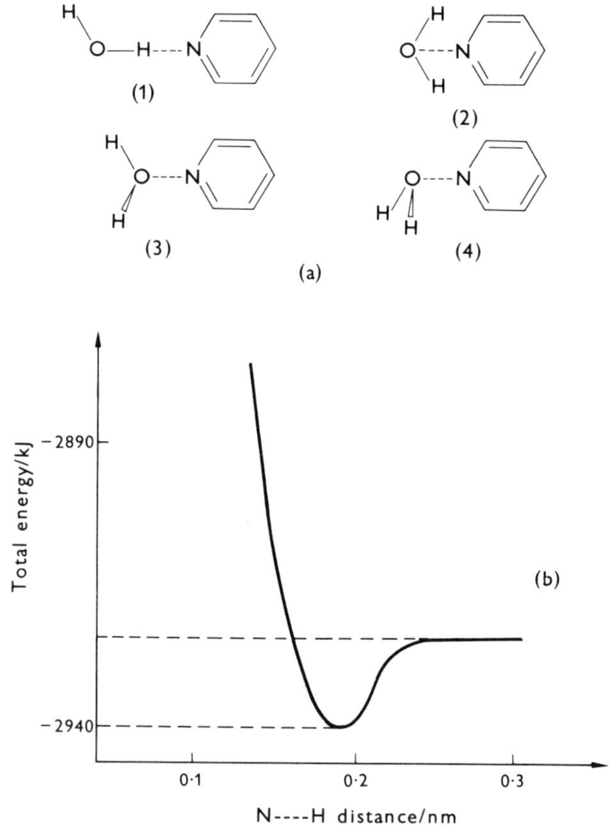

FIG. 7.4. (a) Possible configurations for the H-bonded H_2O-pyridine adduct; (b) total energy – calculated by Extended Hückel Theory – of configuration (1) as a function of r(N-----H). (After S. N. Vinogradov and R. H. Linnel, Reference [3].)

pyridine–water system. An energy minimum was found only for configuration (1) in Fig. 7.4. Configurations (2), (3), and (4) are, of course, unlikely to cause stabilization; an important consequence of such studies, however, is that H-bonding must be directional, as had been known for certain other systems from diffraction studies. H-bonds are always directed towards the lone pairs of the donor atoms, as shown in Fig. 7.5 for liquid HF and for ice.

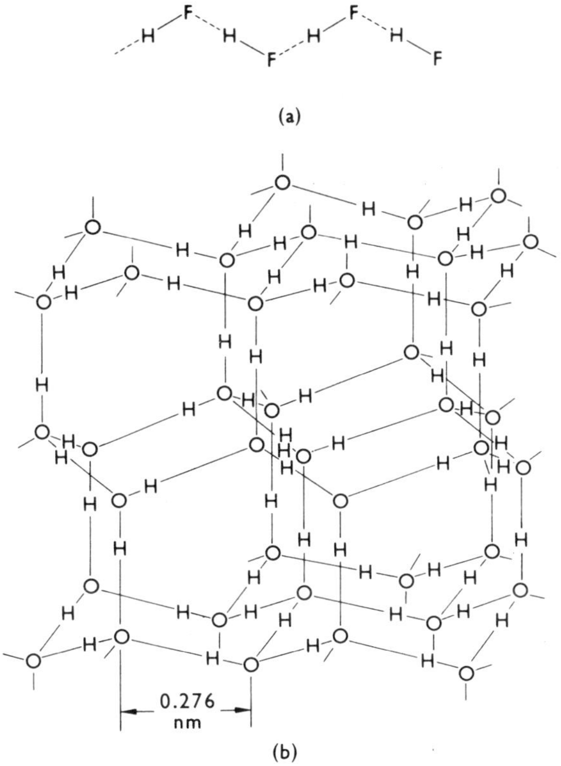

FIG. 7.5. Directional H-bonding in (a) liquid hydrogen fluoride and (b) the ordinary hexagonal form of ice. (From Reference [3].)

7.1.4 Water

The most important solvent in inorganic chemistry is water. Water is an intermolecularly H-bonded liquid and is unusual in many respects; it is a highly structured liquid (see Fig. 7.5) and has a high dielectric constant. It is also unusual in undergoing self-ionization, which shows that it possesses

$$H_2O(l) + H_2O(l) \rightleftharpoons H_3O^+ + OH^-; \quad K_{eq} \approx 10^{-14}\ mol^2\ dm^{-6} \text{ at } 298\ K.$$

both acidic and basic properties. Note that the value of K_{eq} is valid provided that we define the activity of $H_2O(l)$ to be unity. The convenient liquid range and high degree of purity attainable have provided inorganic chemists with a very attractive, but extremely complex system for study.

There are several models of the structure of liquid water. Recent work has suggested that the 'network/cluster' model may be the most acceptable. This suggests that the liquid consists of 'clusters' of H_2O molecules such that interior molecules of a cluster are H-bonded to four others; exterior molecules are bonded to one, two, or three others. Between clusters there are free molecules and the process of interchange of free and H-bonded water occurs in about 10^{-11} seconds.

On dissolving an ionic substance in water, the water molecules become strongly orientated about the ions owing to ion-dipole forces. This ordering is greater than that in the pure solvent but usually extends to a depth no greater than a few layers of H_2O molecules (the hydration shell). This is the 'structure making' region. Beyond the hydration shell, the diminished charge of the ion is still sufficient to disrupt the structure of liquid water into a disordered form. This is the 'structure breaking' region. Beyond this region the water is regarded as being normal.

The concept of structure breaking and making explains the following facts:

(1) Ions with a large charge: radius ratio (e.g. Li^+, Na^+, Mg^{2+}, Al^{3+}, F^-) give a nett negative contribution to the excess entropy of solution of an ionic solute.

(2) Ions with a small charge: radius ratio (e.g. K^+, Cs^+, I^-, ClO_4^-) have a thin hydration sphere (one H_2O in depth) with a deeper surrounding layer of disordered molecules. These ions contribute a nett positive excess entropy of solution.

7.2 **Solvation** [4, 5]

7.2.1 The Nature of Solvation

Solvation involves the donation of an electron pair from a solvent molecule to an electron acceptor. In inorganic chemistry the cation, rather than the anion, is preferentially solvated.

Nuclear magnetic resonance (nmr) data have given very useful insights into the nature of solvation since the position of an nmr peak depends upon the chemical environment of the resonating nucleus. Separate peaks for bound and free solvent are observed if a solvated metal ion undergoes slow solvent exchange. This is the case for solutions containing cations of high charge: radius ratio and solvent molecules of high polarity. Other combinations of metal ion and solvent can give rise to conditions of rapid solvent exchange which yield a single, averaged nmr peak. For example, Al^{3+}/dimethyl sulphoxide solutions give separate nmr peaks but Co^{2+}/methyl cyanide solutions yield an averaged peak.

Therefore, solvation is a dynamic process related to the strength of the metal–solvent interaction. The enthalpy of solvation of metal ions is an important quantity since it is responsible for the degree of solubility of many salts; a discussion of this quantity follows in the next section.

7.2.2 Solubilities of Salts in Water

The dissolution of an ionic compound may be represented by

$$M_nN_m(s) + aq = nM^{m+}(aq) + mN^{n-}(aq) \qquad (7.1)$$

The free energy charge of solution, $\Delta_s^{aq}G^{\ominus}$ is given by

$$\Delta_s^{aq}G^{\ominus}(M_nN_m, s) = -RT \ln \ \{M^{m+}\}^n \ \{N^{n-}\}^m \qquad (7.2)$$

where the quantities in parentheses { } are activities. In dilute solutions, these may be replaced by concentrations and, under these conditions, the solubility of a compound is often expressed by the solubility product, K_{sp};

$$K_{sp} = [M^{m+}]^n \ [N^{n-}]^m \qquad (7.3)$$

$\Delta_s^{aq}G^{\ominus}$ varies with temperature according to the equation (7.4)

$$d(\Delta_s^{aq}G^{\ominus}/T)/dT = -\Delta_s^{aq}H^{\ominus}/T^2 \qquad (7.4)$$

where $\Delta_s^{aq}H^{\ominus}$ is the enthalpy of solution of the solid compound. $\Delta_s^{aq}H^{\ominus}$ is often positive and, therefore, the solubility of a salt should increase with temperature. The solubilities of hydrated salts increase more sharply with temperature than those of anhydrous and less hydrated salts, as $\Delta_s^{aq}H^{\ominus}$ must include the enthalpy change for such dehydration reactions as

$$M_nN_m \, . xH_2O(s) = M_nN_m(s) + xH_2O(l)$$

Such enthalpy changes are always positive. We will now examine the factors which contribute to the enthalpies and entropies of solution of inorganic salts.

Enthalpy Changes

Figure 7.6 summarizes the enthalpy changes which occur on dissolution of an ionic solute in water. If entropy changes are relatively unimportant, $\Delta_s^{aq}H^{\ominus}$ must be negative for the salt to be soluble. Therefore, the enthalpy of hydration ($\Delta_g^{aq}H^{\ominus}$) of the gaseous ions must exceed the sum of the lattice energy (U) and, where appropriate, the enthalpy of dehydration ($\Delta^dH^{\ominus}$) of the compound.

Since any compound contains at least two ions of opposite charge, it is not possible to measure the absolute value of the individual $\Delta_g^{aq}H^{\ominus}$ values of the cations and anions. We therefore define a zero, $\Delta_g^{aq}H^{\ominus}(H^+, g)$ such that all other values of $\Delta_g^{aq}H^{\ominus}$ are relative to it. Some relative $\Delta_g^{aq}H^{\ominus}$ values for common cations and anions are listed in Table 7.2. Small ions of high charge

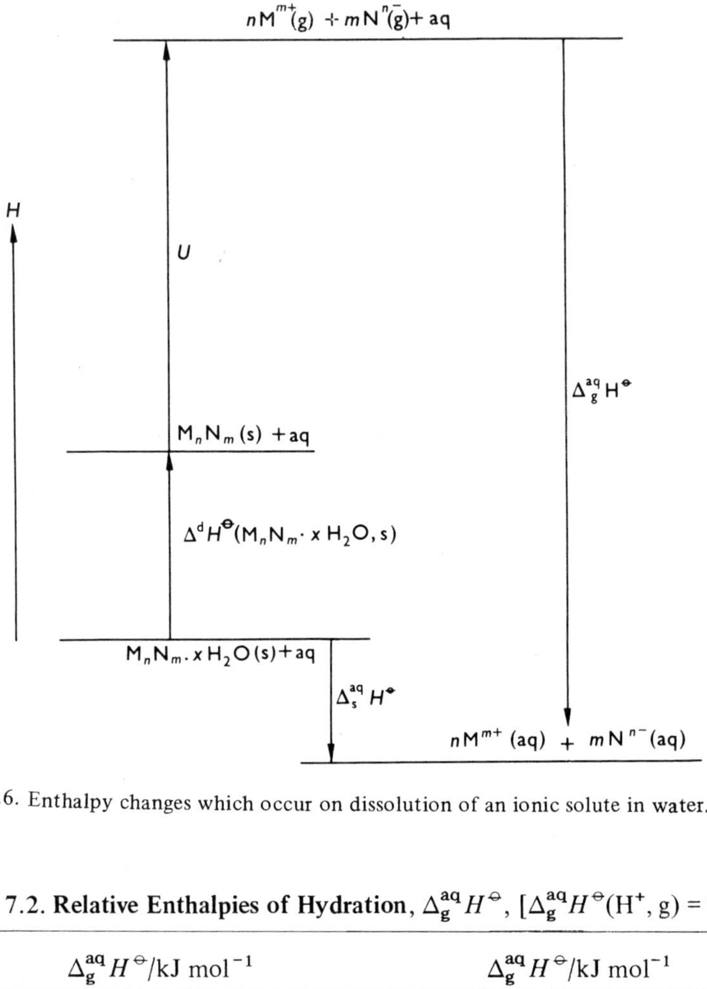

FIG. 7.6. Enthalpy changes which occur on dissolution of an ionic solute in water.

Table 7.2. Relative Enthalpies of Hydration, $\Delta_g^{aq}H^\ominus$, $[\Delta_g^{aq}H^\ominus(H^+, g) = 0]$

	$\Delta_g^{aq}H^\ominus$/kJ mol^{-1}		$\Delta_g^{aq}H^\ominus$/kJ mol^{-1}
H^+	0·0	Cu^{2+}	81·2
Li^+	576	Zn^{2+}	137
Na^+	685	Sc^{3+}	−688
K^+	770	La^{3+}	−10·5
Be^{2+}	−305	Al^{3+}	−1387
Mg^{2+}	259	Ga^{3+}	−1412
Ca^{2+}	590	In^{3+}	−836
Cr^{2+}	332	F^-	−1532
Co^{2+}	127	Cl^-	−1459
Ni^{2+}	75·7	Br^-	−1425

have the larger values of $\Delta_g^{aq}H^\ominus$. There is no *simple* quantitative correlation between the ionic charge and radius and $\Delta_g^{aq}H^\ominus$. A simple theoretical model of moderate accuracy is, however, available.

In Chapter 1, we discussed the Born equation for the free energy change on placing spherical ions into a solvent of dielectric constant, ϵ. There is also an equation for the entropy change accompanying this process. The equations should be applicable to the theoretical calculation of $\Delta_g^{aq}G^\ominus$, $\Delta_g^{aq}S^\ominus$ and, therefore, $\Delta_g^{aq}H^\ominus$ for ionic compounds:

$$\Delta_g^{aq}G^\ominus = -\frac{Ne^2z^2}{8\pi r\epsilon_0}\left(1-\frac{1}{\epsilon}\right) = -\frac{67{\cdot}8z^2}{r}\ \text{kJ mol}^{-1}$$

$$\Delta_s^{aq}S^\ominus = \frac{Ne^2z^2}{8\pi\epsilon r\epsilon_0}\left(\frac{\partial \ln \epsilon}{\partial T}\right)_P = -\frac{3{\cdot}94z^2}{r}\ \text{J K}^{-1}\ \text{mol}^{-1}$$

$$\Delta_g^{aq}H^\ominus = -69z^2/r\ \text{kJ mol}^{-1}$$

In these equations, z is the charge on the ion of radius r/nm and ϵ_0 is the permittivity of free space.

Table 7.3. Enthalpies of Hydration ($\Delta_g^{aq}H^\ominus$) of Gaseous Ions; Lattice Energies (U); Changes in Entropy, $T\Delta_s^{aq}S^\ominus$, Free Energy, $\Delta_s^{aq}G^\ominus$, and Enthalpy, $\Delta_s^{aq}H^\ominus$, for the Solution of some Solid Ionic Compounds

	$-\Delta_g^{aq}H^\ominus$ (Exp.)	$-\Delta_g^{aq}H^\ominus$ (Born Eq.)	U	$\Delta_s^{aq}G^\ominus$	$\Delta_s^{aq}H^\ominus$	$T\Delta_s^{aq}S^\ominus$
			kJ mol^{-1}			
LiF	1004	957	1008	+13·8	+ 4·2	− 9·6
LiCl	866	825	828	+41·4	−37·6	+ 3·8
NaCl	770	767	774	− 9·2	+ 3·8	+13·0
KCl	682	697	699	− 6·7	+16·7	+23·4
AgF	974	850	954	−14·6	−20·5	− 5·8
AgCl	838	718	904	+55·6	+65·6	+10·0
AgBr	817	698	887	+70·2	+84·5	+14·2
CaF_2	2598	2529	2611	+56·1	+13·4	−42·7
$CaCl_2$	2330	2267	2247	−65·2	−82·8	−17·6
AlF_3	6160	6077	6140	+18·8	−134	−115
$AlCl_3$	5770	5682	5460	−24·6	−328	−303
$Na(NO_3)$	721	748*	741	− 6·7	+20·5	+26·8
$Ca(NO_3)_2$	2232	2230*	2213	−32·2	−18·9	+13·3

* using Kapustinskii radius of NO_3^-.

The equations are only approximations and there are several theoretical objections to their precise applicability. For example, in order to obtain good agreement with experimental values of $\Delta_g^{aq} H^\ominus$ for cations, an empirical correction must be made:

$$r(\text{cation})/\text{nm} = r(\text{PC})/\text{nm} + 0{\cdot}085/\text{nm}$$

r(PC) is the Pauling Crystal Radius of the cation; 0·085 nm is the radius of the oxygen atom in water and allows for the fact that the oxygen end of the H_2O molecule cannot approach a cation as closely as the hydrogen end can approach an anion. No correction is applied for anions.

The enthalpy of hydration of a gaseous ionic compound can, therefore, be estimated to be

$$\Delta_g^{aq} H^\ominus(\text{M}_n^{m+}\ \text{N}_m^{n-}, \text{g}) = -69 \{ [n(m+)^2/(r(\text{PC}) + 0{\cdot}085)] + m(n-)^2/r_\text{A} \} \quad (7.5)$$

where r_A is the ionic radius of the anion in nm. In Table 7.3 are listed some experimental and calculated values of $\Delta_g^{aq} H^\ominus$ together with the experimentally determined values of $\Delta_s^{aq} G^\ominus$, $\Delta_s^{aq} H^\ominus$, $T\Delta_s^{aq} S^\ominus$, and U.

The values of $\Delta_g^{aq} H$ calculated from the Born equation certainly give poor absolute agreement with theory; however, the predicted trends are good and, for many compounds, there is good quantitative agreement. Therefore, the Born equations can be used with confidence to predict the thermodynamic changes on hydration (or solvation) but in certain cases we must expect there to be discrepancies. This seems to be particularly likely when the compound being considered is appreciably covalent – for example, in the case of the silver halides. The Born model is for *ionic* compounds and breaks down in the same way as the Born–Landé equation (3.8) for covalent compounds.

From the free energies and enthalpies of solution (Table 7.3), it is clear that the most important factor in determining solubility is the relative magnitude of $\Delta_g^{aq} H^\ominus$ compared with U. This is not true for all compounds. For example, both sodium nitrate and calcium nitrate are more soluble, as judged by $\Delta_s^{aq} G^\ominus$, than we would expect on pure enthalpy grounds. This is due to the very favourable positive entropy changes for these compounds.

Entropy Changes

Approximate quantitative calculations of $\Delta_g^{aq} S^\ominus$ may be made with the help of the Born equations. There are, however, some qualitative points which can be discussed with reference to the expected magnitude of $\Delta_s^{aq} S^\ominus$.

From Table 7.3 we see that compounds containing large ions of low charge tend to have the more positive $T\Delta_s^{aq} S^\ominus$. These compounds, such as nitrates, tend to be more soluble so long as $\Delta_s^{aq} H^\ominus$ is not too unfavourable.

Since $T\Delta_s^{aq}S^\ominus$ is positive, $-\Delta_s^{aq}G^\ominus$ (and hence their solubility) increases with temperature.

The removal of ions from a crystal lattice involves the following entropy changes:

(1) An increase in translational and (where appropriate) rotational entropy as the ions pass into solution.

(2) A change in entropy of the solvent (see 7.1.4)

(3) An increase in local solvent order about the metal ions, causing an entropy decrease.

For ions of low charge and large size, (1) and (2) increase $\Delta_s^{aq}S^\ominus$ and may cause this term to be dominant. Factor (3) is important for small, highly charged ions and $\Delta_s^{aq}S^\ominus$ is, as a consequence, unfavourable for compounds which contain them.

7.2.3 Non-Aqueous Solvents [6, 7]

Water is a very common solvent but, owing to its high polarity and strong solvating ability, many compounds undergo solvolytic reactions and cannot be isolated in aqueous solutions. Non-aqueous solvents have considerable utility, permitting reactions to proceed which could not take place in water. For example, metal alcoholates are precipitated on mixing a potassium alcoholate with a metal salt in liquid ammonia:

$$M^{2+} + 2KOR \xrightarrow[NH_3]{liq} M(OR)_2 + 2K^+$$

The effect of basic and acidic solvents on acidic and basic properties has already been mentioned briefly in Chapter 3. In this chapter we are mainly concerned with models describing the states of solutes in non-aqueous solutions. The two that we shall discuss are:

(a) The Solvent System Model
(b) The Coordination Model

According to model (a) an acid is defined as a substance which gives rise to a cation characteristic of the solvent; a base is a substance which gives rise to an anion characteristic of the solvent. Inherent in this model, therefore, is the concept of autoionization such as

$$2H_2O \rightleftharpoons H_3O^+ + OH^-$$

$$2NH_3 \rightleftharpoons NH_4^+ + NH_2^-$$

Thus NH_4Cl is an acid, and $NaNH_2$ a base, in liquid ammonia. Whilst this model is useful in a qualitative fashion it has the following disadvantages:

(1) Some solvents, such as sulphur dioxide, do not autoionize to any extent but acid-base neutralizations can be carried out in them.

(2) The model does not account for the properties of non-ionizing solvents.

(3) No discussion of thermodynamic properties is possible.

The more recent Coordination Model, suggested by Drago and Purcell, overcomes many of these difficulties. This model focuses attention on the strength of the interactions between solute and solvent molecules. Aspects of these interactions have been discussed in Chapter 1 in terms of soft and hard acid and base theory.

There is ample evidence for the formation of complex ionic species involving coordinate bonds to a metal ion. Solvent coordination also occurs in organic systems, as evidenced by the formation of molecular complexes between strong donor and strong acceptor molecules.

Therefore, we shall assume that the predominant forces between all solute and solvent molecules are as a result of coordinate bond formation. Drago and Purcell extended this concept and stated that: 'the donor (or acceptor) strength of a solvent are the most significant properties that, to a first approximation, determine the nature of solute species in solution'. Most of the discussion that follows is concerned with simple metallic salts as solutes but the concepts are applicable to most *simple* systems.

If a metal salt, MX_n, dissolves in a donor solvent, S, an addition compound is the first solute species to be formed:

$$MX_n + pS = MS_pX_n$$

The next steps will depend on the strength of the M–S interaction: If the coordination number of the metal in the addition compound is maintained in the products, reactions of the following type proceed in basic solvents:

$$MS_pX_n + mS \rightleftharpoons MS_{p+m}X_{n-m} + mX^-$$

If the coordination number of the metal is variable, then anionic species of the type $MS_iX_j^{(n-j)}$, in which $(i + j)$ is less than $(p + n)$ may form. Therefore, if the coordination model is correct similar solute species are to be expected in non-aqueous solvents if the solvents are of similar donor strength. The solvent system predicts that different solute species would be formed. To prove the correctness of the coordination model, consider the behaviour of $FeCl_3$ in the non-aqueous solvents phosphorous oxychloride, $POCl_3$, and triethylphosphate, $PO(OEt)_3$. Spectroscopic studies suggest the presence of the $FeCl_4^-$ ion in each system and this is confirmed by the ability to titrate $FeCl_3$ with Cl^- in each solvent. The end point corresponds to a 1:1 molar ratio of reactants. According to the solvent system, $FeCl_3$ is an acid in either solvent:

$$2POCl_3 \rightleftharpoons POCl_2^+ + POCl_4^-$$
$$POCl_3 + FeCl_3 \rightarrow POCl_2^+ + FeCl_4^-$$

$$2PO(OEt)_3 \rightleftharpoons PO(OEt)_2^+ + PO(OEt)_4^-$$
$$PO(OEt)_3 + FeCl_3 \rightarrow PO(OEt)_2^+ + FeCl_3(OEt)^-$$

Therefore, only in $POCl_3$ would we expect the formation of $FeCl_4^-$. The coordination model, however, correctly explains the formation of this ion in either solvent:

$$POR_3 + FeCl_3 \rightleftharpoons FeCl_3(POR_3)_n \tag{7.6}$$
$$POR_3 + FeCl_3(POR_3)_n \rightleftharpoons FeCl_2(POR_3)_{n+1} + Cl^- \tag{7.7}$$
$$FeCl_3(POR_3)_n + Cl^- \rightleftharpoons FeCl_4^- + n(POR_3) \tag{7.8}$$

Such a reaction sequence with R = Cl or OEt and n = 1 accounts for all experimental observations on this system. The extent to which such reactions as (7.6–7.8) proceed depends on the donor strength of the solvent. Such

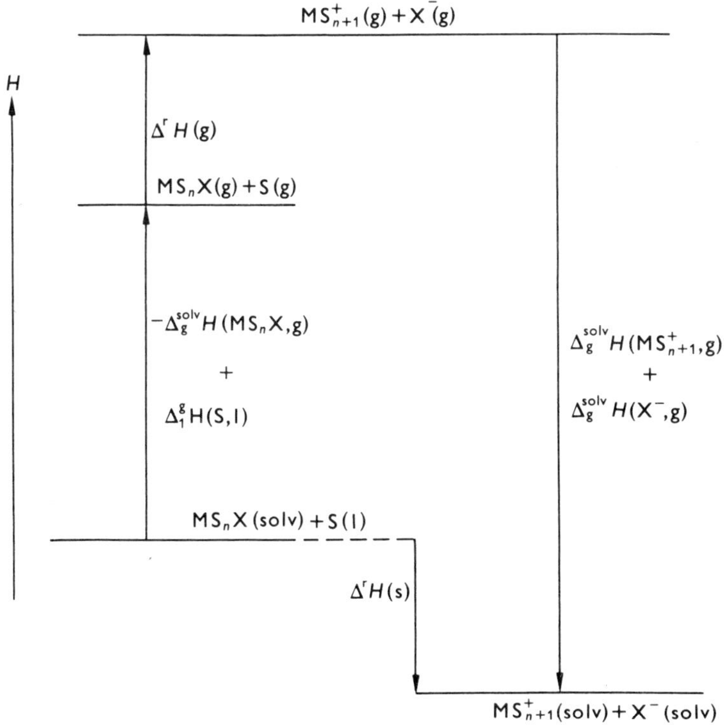

FIG. 7.7. Enthalpy changes for equation (7.9).

solvents are called *ionizing solvents.* We now examine the thermochemical relationships inherent in this model.

For the simple case of a polar, univalent, covalent compound MX, the adduct first formed in the solvent S will be MS_nX. The ultimate products are defined by (7.9).

$$MS_nX\,(solv) + S \rightarrow MS^+_{n+1}\,(solv) + X^-\,(solv) \qquad (7.9)$$

The thermochemical cycle in Fig. 7.7 summarizes the enthalpy changes in (7.9); $\Delta^{solv}_{g}H(MS_nX, g)$ is the enthalpy of solvation of MS_nX; $\Delta^{g}_{l}H(S, l)$ is the enthalpy of vaporization of S(l); $\Delta^{r}H(s)$ and $\Delta^{r}H(g)$ are the solvent and gas-phase enthalpy changes for reactions analogous to (7.9).

Similar cycles can be constructed for multivalent compounds. The most important factor responsible for the formation of ionic species in solution is the gas-phase enthalpy change, $\Delta^{r}H(g)$. This cannot be calculated since it is only possible to measure the corresponding solution phase enthalpy change. The two enthalpy changes do not necessarily parallel each other because of the large, and unpredictable values of the other enthalpy changes. The largest of these are the solvation terms. Closer examination of this cycle reveals that the formation of ionic species is a reaction involving the competition of S and X^- for M^+. Therefore, a simpler approach to the problem is to measure the solvating ability of S with reference to a fixed standard. The result would then be a series of relative solvent donor strengths. A suitable measure of such donor strengths is the enthalpy change, $\Delta H(7.10)$, of the following reaction carried out in a weakly coordinating solvent (for example CCl_4):

$$A(solv) + B(solv) \rightleftharpoons AB(solv) \qquad (7.10)$$

This enthalpy change closely parallels the corresponding gas-phase enthalpy change. In addition to direct calorimetric measurement, the measurement of equilibrium constant variation with temperature and spectroscopic results have been useful. Spectroscopic measurements are faster and more simple to make than calorimetric measurements. A very useful infra-red method involves the relationship between the difference in 'frequency' (ν/cm^{-1}) for the O–H stretching vibration of free and complexed phenol and $\Delta H(7.10)$ where B is phenol:

$$-\Delta H(7.10)/kJ\ mol^{-1} = 0{\cdot}067/cm^{-1} + 2{\cdot}636$$

Some values of $\Delta H(7.10)$ are listed in Table 7.4 for iodine and phenol acceptors.

In Chapter 1 we discussed the 'hard and soft' properties of acids and bases. It is important, therefore, when using thermodynamic data (such as those in Table 7.4) to predict the donor strength of a solvent, that a similar acceptor molecule is used as the reference. In Table 7.4 phenol can be regarded as a

Table 7.4. Values of $-\Delta H$ (7.9) for Various Donors with Iodine or Phenol in Weakly Coordinating Solvents

	$-\Delta H(7.9)$/KJ mol^{-1}	
Donor	Iodine	Phenol
$(CH_3)_2SO$	18·4	27·2
$HC(O)N(CH_3)_2$	15·5	25·5
$CH_3C(O)N(CH_3)_2$	16·7	26·8
$CH_3C(S)N(CH_3)_2$	39·7	23·0
$(C_2H_5)_2O$	17·6	20·9
$(C_2H_5)_2S$	32·6	19·2
C_5H_5N	32·6	33·8
NH_3	20·1	34·3
$(CH_3)_3N$	50·6	39·6
CH_3NO_2	–	7·9

hard acid and iodine as a soft acid, so several useful correlations should be possible. We now illustrate the application of the coordination model to some chemical systems.

(a) *The Formation and Stability of Ionic Halo Complexes*

The order of stability of $CoCl_4^{2-}$ in a number of solvents is:

$$CH_2Cl_2 > CH_3NO_2 > CH_2C(O)N(CH_3)_2 \sim HC(O)N(CH_3)_2 > (CH_3)_2SO$$

This sequence represents the ease of formation from $CoCl_4^{2-}$ of species in which chlorine atoms are replaced by solvent to form $CoCl_3S^-$, CoS_5Cl^+, and CoS_6^{2+}. This sequence closely parallels the increasing enthalpy of adduct formation of the solvents with phenol (Table 7.4) which is a σ-bonding acceptor.

(b) *Acidic and Basic Properties*

Solutions of very strong donor solutes in the solvents $CHCl_3$, CH_2Cl_2, $(CH_3)_2SO$ (which are very weak acceptors) form only hydrogen bonded adducts as the main products. Solvents of greater acceptor power easily form adducts with relatively weak donors. For example, in water:

$$B: + H{-}O{-}H \rightarrow [B:\cdots\cdots H{-}O{-}H]$$

A second reaction occurs in which OH^- is displaced by B

$$[B:\cdots\cdots H{-}O{-}H] \rightleftharpoons BH^+ + OH^-$$

The extent to which this reaction proceeds depends on the donor strength of B. If B is a much stronger donor than OH^- (e.g. B = NH_2^-, OR^-, H^-) then the basic strength as measured by $[OH^-]$ is independent of the nature of the base. Therefore, water is said to be a *levelling* solvent for bases.

A similar argument holds for strong acids. Water is also a levelling solvent for acids in that all strong acids appear to have the same strength in water. For the simple acid denoted by HY, the initial adduct will be

$$\left[\begin{matrix} H \\ H \end{matrix} \!\!> O \cdots\cdots H \text{———} Y \right]$$

Acid strength might be expected to increase with the acceptor strength of Y. However, if Y is a weaker donor than H_2O, then the second reaction occurs

$$\left[\begin{matrix} H \\ H \end{matrix} \!\!> O \cdots\cdots H \text{———} Y \right] \rightarrow H_3O^+ + Y^-$$

In other words, the strongest acid in water is H_3O^+ and the strongest base is OH^-. This concept is easily extended to other solvents – strongly basic solvents have a levelling effect on acids; strongly acidic (i.e. acceptor) solvents have a levelling effect on bases. Some examples are listed in Table 7.5.

Table 7.5. Examples of Non-Aqueous Solvents which Undergo Ionization

Solvent	Remarks	Strongest Acid	Strongest Base
NH_3(l)	Strong Donor	NH_4^+	NH_2^-
CH_3CO_2H(l)	Strong Acceptor	$CH_3CO_2H_2^+$	$CH_3CO_2^-$
HCl(l)	Strong Acceptor	H^+	HCl_2^-
HF(l)	Strong Acceptor	H^+	HF_2^-
H_2SO_4(l)	Very Strong Acceptor	$H_3SO_4^+$	HSO_4^-
$POCl_3$(l)	Donor/Acceptor	$POCl_2^+$	Cl^-
NOCl(l)	Donor/Acceptor	$NO(NOCl)^+$	Cl^-
N_2O_4(l)	Donor/Acceptor	NO^+	NO_3^-

Many reactions in non-aqueous solvents can be rationalized in terms of the coordination model described in this section. These include:

$$M^+H^- + NH_3 \rightarrow M^+ + NH_2^- + H_2 \text{ (Displacement of } NH_2^-\text{)}$$

$$BCl_3 + 6NH_3 \rightarrow B(NH_2)_3 + 3NH_4Cl \text{ (Displacement of } H^+\text{)}$$

$$\left.\begin{matrix} M^+Cl^- + H_2SO_4 \rightarrow M^+ + HSO_4^- + HCl \\ H_3PO_4 + H_2SO_4 \rightarrow H_4PO_4^+ + HSO_4^- \end{matrix}\right\} \text{ (Displacement of } HSO_4^-\text{)}$$

Other examples will be found by consulting the references at the end of this chapter.

7.2.4 Electrode Potentials [8, 9]

The standard electrode potential, $E^{\ominus}$, of a *reduction* process is related to the standard free energy of that process. For the reaction, (7.11), in which OX the free energy change is given by

$$OX + ne \rightarrow RED \qquad (7.11)$$

reduced species and n is the number of electrons, e, involved in the reduction, the free energy change is given by

$$\Delta G^{\ominus} = -nFE^{\ominus}$$

where $\Delta G^{\ominus}$ is in J mol^{-1} if $E^{\ominus}$ is in volts and F is the Faraday Constant ($9{\cdot}6487 \times 10^4$ C mol^{-1}).

The oxidized and reduced species (either or both of which may be in solution) of (7.11) are called a 'redox couple'. Any chemical reaction involving electron transfer must be comprised of two such couples: one for oxidation (loss of electrons) and the other for reduction (gain of electrons). It is clear, therefore, that the $E^{\ominus}$ of a single reaction cannot be assigned an absolute value. It is conventional to refer all values of $E^{\ominus}$ to an arbitrary zero; in aqueous solution, this is defined as $E^{\ominus}$ of the redox reaction:

$$2H^+(aq) + 2e \rightarrow H_2(g) \quad E^{\ominus} = 0{\cdot}0 \text{ volts}$$

In this reaction (which represents the standard hydrogen electrode) H^+(aq) and H_2(g) are at unit activity and fugacity, respectively, and the defined temperature is 298 K.

$E^{\ominus}$ values are conventionally listed as standard potentials at 298 K with all species at standard activities or fugacities. Some examples are listed in Table 7.6. (The $E_1^{\ominus}$ values in this table are discussed later.) For non-standard conditions, the electrode potential of a redox reaction is related to the standard potential by the *Nernst Equation*:

$$E = E^{\ominus} + \frac{RT}{nF} \ln \frac{OX^m}{RED^n}$$

OX and RED are the activities or fugacities of the oxidized and reduced species; m and n are their stoichiometric coefficients in the reduction reaction. For example,

$$\underbrace{MnO_4^- + 8H^+}_{OX} + 5e = \underbrace{Mn^{2+} + 4H_2O}_{RED} \quad E^{\ominus} = 1{\cdot}51 \text{ V}$$

Since the activity of water can be assumed to be near unity in dilute solution, we obtain:

$$E = 1{\cdot}51 + \frac{RT}{nF} \ln \frac{[MnO_4^-]\,[H^+]^8}{[Mn^{2+}]}$$

At 298 K,

$$E = 1{\cdot}51 + 0{\cdot}059 \log_{10} \frac{[MnO_4^-]\ [H^+]^8}{[Mn^{2+}]}$$

In fairly dilute solutions, concentrations can be substituted for activities to yield sufficiently accurate results.

The main usefulness of electrode potentials is in predicting the spontaneity of chemical reactions. If the electrode potential of the reduction process is more positive than that of the oxidation process, then the reaction will proceed spontaneously. Of course, this information gives no clue as to the speed of a reaction. Some reactions may proceed spontaneously but very slowly. For example, Sn(II) appears to be unaffected by hydroxylamine in acid solution, which is surprising since the E° values predict that Sn(II) should be oxidized.

$$Sn^{4+}(aq) + 2e = Sn^{2+}(aq) \quad E^{\ominus} = 0{\cdot}15\ V$$

$$NH_3OH^+(aq) + 2H^+(aq) + 2e = H_2O(l) + NH_4^+(aq) \quad E^{\ominus} = 1{\cdot}35\ V$$

It is sometimes possible to adjust the value of an electrode potential by changes in concentration, temperature, or pressure to permit redox reactions to proceed. The results can be predicted with reasonable accuracy by the Nernst equation.

In this chapter, we are concerned with the underlying reasons for the variation in $E^{\ominus}$ values, rather than their applications. We shall attempt to outline some aspects of the variation in $E^{\ominus}$ for some simple reactions that are amenable to analysis by thermochemical cycles. There are two possible approaches: (1) to calculate $E^{\ominus}$ values entirely from first principles, using tabulated and theoretical data, and to compare the results with experimental values: (2) to subtract from the $E^{\ominus}$ values those thermodynamic quantities which are well defined and to analyse the residual variation in $E^{\ominus}$ by qualitative or semi-quantitative reasoning.

We prefer the second of these approaches since it is rarely possible to attempt a full thermodynamic analysis of electrode reactions. Let us look at an example of this approach for a particularly simple case:

For the series of reactions described by

$$M^{n+}(aq) + ne \rightarrow M(s) + aq. \tag{7.12}$$

where M is a metal, we expect that the entropy changes will be finite but should not vary appreciably for a related series of metals in the same oxidation states. Figure 7.8 illustrates the enthalpy changes for reaction (7.12) and it can be seen that $\Delta_g^{aq}H^{\ominus}(M^{n+}, g)$ is a large and important quantity. Owing to the similar, but less predictable, variation in $\left[\sum_{i=1}^{n} I_i + \Delta_s^g H^{\ominus}(M, s)\right]$ there is

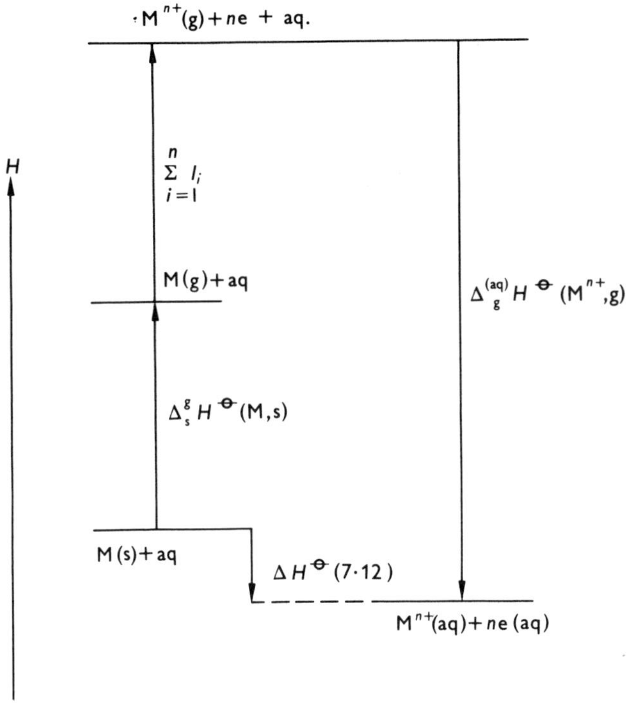

FIG. 7.8. Enthalpy changes for a simple redox reaction (equation (7.12)).

Table 7.6. Electrode Potentials, $E^{\ominus}$, in volts for some Common Redox Couples

Couple	$E^{\ominus}$	$E_1^{\ominus}$	Couple	$E^{\ominus}$	$E_1^{\ominus}$
Li^+, Li	−3·045	−10·06	Ti^{2+}, Ti	−1·63	−14·13
Na^+, Na	−2·714	− 8·96	V^{2+}, V	−1·18	−14·15
K^+, K	−2·925	− 8·16	Cr^{2+}, Cr	−0·90	−14·40
Rb^+, Rb	−2·925	− 7·96	Mn^{2+}, Mn	−1·18	−13·98
Cs^+, Cs	−2·925	− 7·61	Fe^{2+}, Fe	−0·44	−14·44
			Co^{2+}, Co	−0·28	−14·73
Be^{2+}, Be	−1·85	−17·23	Ni^{2+}, Ni	−0·25	−15·12
Mg^{2+}, Mg	−2·37	−14·41	Cu^{2+}, Cu	+0·34	−15·18
Ca^{2+}, Ca	−2·87	−12·76	Zn^{2+}, Zn	−0·76	−14·92
Sr^{2+}, Sr	−2·89	−12·03	Al^{3+}, Al	−1·67	−20·47
Ba^{2+}, Ba	−2·90	−11·41	Ga^{3+}, Ga	−0·52	−20·52
Ra^{2+}, Ra	−2·92	−11·48	In^{3+}, In	−0·34	−18·66
			Tl^{3+}, Tl	+0·72	−18·60

no easily identifiable trend in $E^{\ominus}$, as seen in Table 7.6.

Fortunately, both $\sum_{i=1}^{n} I_i$ and $\Delta_s^g H^{\ominus}(M, s)$ are known for many metals so that, with the assumption of constant entropy changes, we can define a new parameter:

$$E_1^{\ominus} = E^{\ominus} - \left[\sum_{i=1}^{n} I_i + \Delta_s^g H^{\ominus}(M, s)/nF\right] \qquad (7.13)$$

This second term in (7.13) removes the irregular influences and leaves the quantity $E_1^{\ominus}$ which reflects *only* the changes in $\Delta_g^{aq} H^{\ominus}(M^{n+}, g)$. The variation in this quantity is clearly seen (Table 7.6) for the following groups of elements:

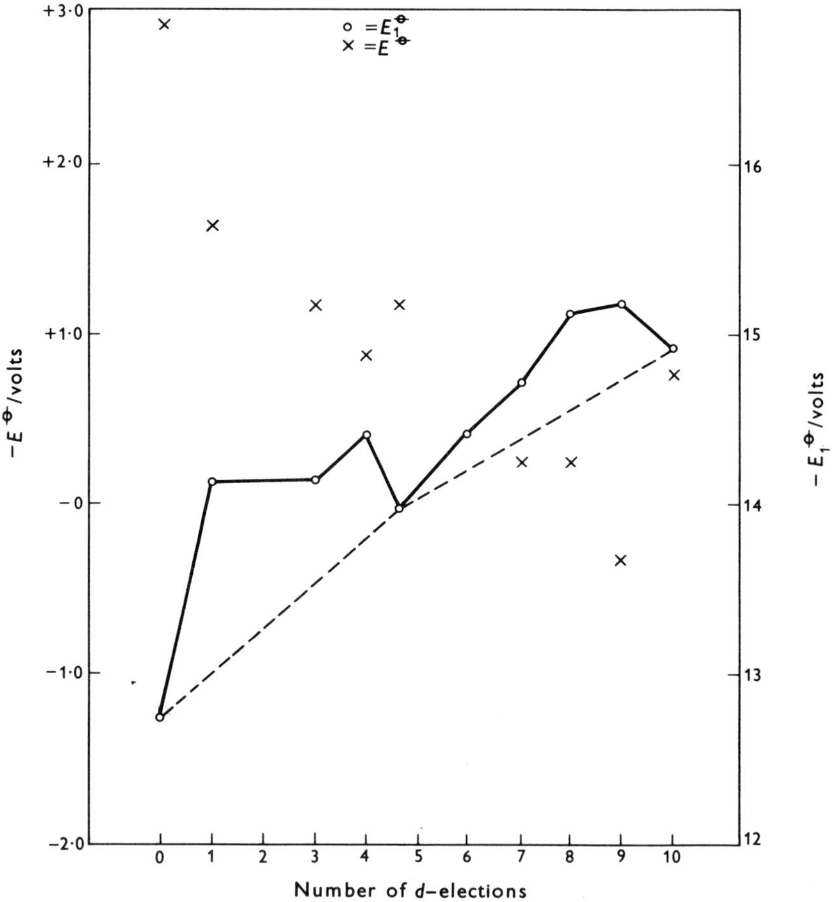

FIG. 7.9. Electrode potentials, $E^{\ominus}$ and $E_1^{\ominus}$ (calculated from equation (7.13) for divalent transition elements.

Non-transition Elements

Within a group, there is a smooth decrease in $E_1^\ominus$ which reflects $\Delta_g^{aq} H^\ominus(M^{n+}, g)$. For the alkali metals, the near constancy of $E^\ominus$ from sodium to caesium shows that the variation in $\Delta_g^{aq} H^\ominus(M^+, g)$ is nearly balanced by that in $[I_1 + \Delta_s^g H^\ominus(M, s)]$.

The similarity of $E_1^\ominus(Al^{3+}/Al)$ and $E_1^\ominus(Ga^{3+}/Ga)$ is caused by the intervention of the 3*d* elements prior to gallium; the increased nuclear charge is not balanced by electron screening. Likewise, $E_1^\ominus(In^{3+}/In)$ and $E_1^\ominus(Tl^{3+}/Tl)$ are similar owing to the intervention of the 4*f* (lanthanide) series before Tl.

The 3d-Transition Elements

On passing from Ca^{2+} to Zn^{2+} there is no recognizable trend in $E^\ominus$; $E_1^\ominus$, however, has a regular periodic variation as shown in Fig. 7.9. This reflects the variation in crystal field stabilization energy of the M^{2+} ions (see Section 4.2.2).

A similar thermochemical analysis can be attempted for anions but with lower accuracy, owing to our less precise knowledge of electron affinities.

REFERENCES

Liquid Structure

[1] Barker, J. A., *Lattice Theories of the Liquid State*, Pergamon, London (1963).

[2] Rowlinson, J. S., *Liquids and Liquid Mixtures*, Butterworths, London (1969).

Hydrogen Bonding and Water

[3] Vinogradov, S. N., and Linnell, R. H., *Hydrogen Bonding,* Van Nostrand Reinhold, New York (1971).

Solvation and Solubility

[4] Phillips, C. S. G., and Williams, R. J. P., *Inorganic Chemistry; Part I*, Oxford University Press, Oxford (1965).

[5] Coetzee, J. F., and Ritchie, C. D., *Solute–Solvent Interactions*, Marcel Dekker, New York (1969).

Non-Aqueous Solvents

[6] Jander, J., and Lafrenz, Ch., *Ionizing Solvents*, Wiley, London (1970).

[7] Drago, R. S., and Purcell, K. F., *Progress in Inorganic Chemistry*, Vol. 6, (Ed. F. A. Cotton), Interscience, New York (1964), p. 271 ff.

Electrode Potentials

[8] Denaro, A. R., *Elementary Electrochemistry*, 2nd edn, Butterworths, London (1971).

[9] Figgis, B. N., *Introduction to Ligand Fields*, Interscience, New York, (1966) p. 271 ff.

Author Index

For each author, the reference numbers in brackets are followed by the numbers of the pages on which the reference number occurs. The symbol (F) indicates a reference to further reading. Bibliographic details of each reference are given at the end of the appropriate chapter.

Subject Index

$ 18.00
4/7/81
Madison